MW01171339

Faces of Pilot Mountain

Intriguing Journey into the Deep Mysteries

J. P. McKelvey

Kelvimages Publishing, LLC
Hillsborough, North Carolina

Typeset in Helvetica and Jenson by TIPS Publishing Services, Carrboro, NC
Cover design by Robert Kern

Library of Congress Cataloging in Publication Data

Faces of Pilot Mountain: Intriguing Journey into the Deep Mysteries / J. P. McKelvey
p. cm.
Includes bibliographical and index references.
Summary: Account of the 1.6-billion-year history of Pilot Mountain in North Carolina found in Surry County between Mt. Airy and Winston-Salem.

ISBN 978-1-890586-81-2 (cloth), 978-1-890586-79-9 (paperback), 978-1-890586-80-5 (ebook)

Printed and bound in the United States.

The author can be contacted at: jackmac2013@gmail.com
Online homepage at http://www.facesofpilotmountain.com

This book is dedicated to author and international speaker, Art Fettig (1929–2021). Art mentored me during our weekly Wednesday morning breakfast which lasted for the duration of the research and writing of this book. Many in the area of Pilot Mountain State Park might also know him from his role as parade marshall of "Mayberry Days" or his stand-up act, "Almost Andy." Art was a delightful man who dedicated much of his life to improving railroad safety.

Strength

Humans Seeking Balance

Unity

Bronze plaque found at the entrance of the trail to the Little Pinnacle at Pilot Mountain State Park

Looking south at Pilot Mountain State Park

CONTENTS

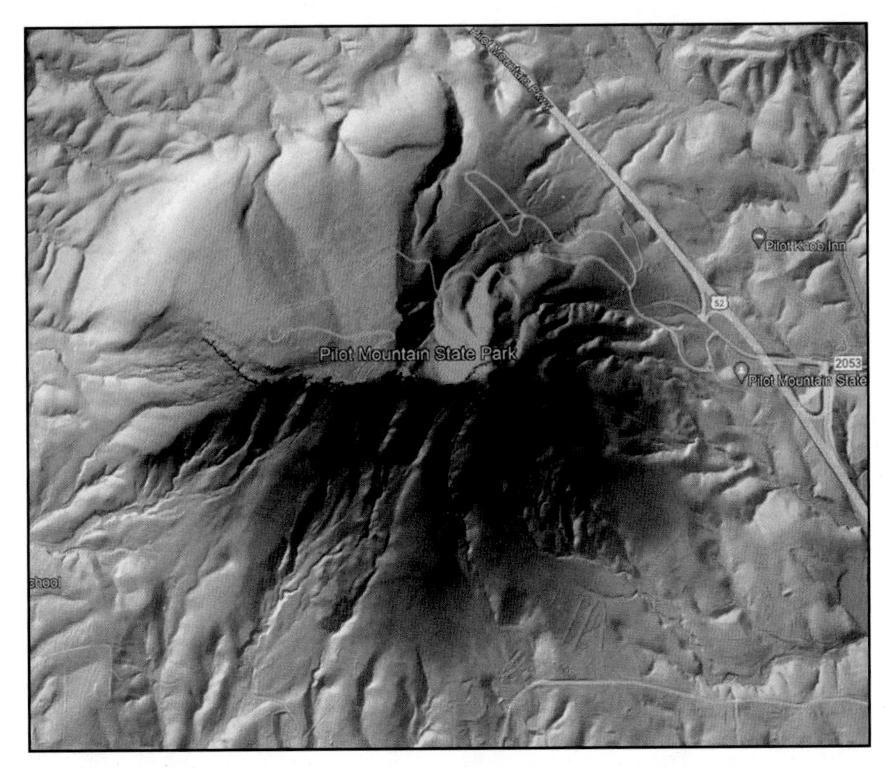

A Laser Imaging, Detection, and Ranging (LIDAR) image of Pilot Mountain State Park. LIDAR software removes the foliage so you only see the ground (courtesy of North Carolina Division of Emergency Management with geological software enhancement by Matthew Hospodar).

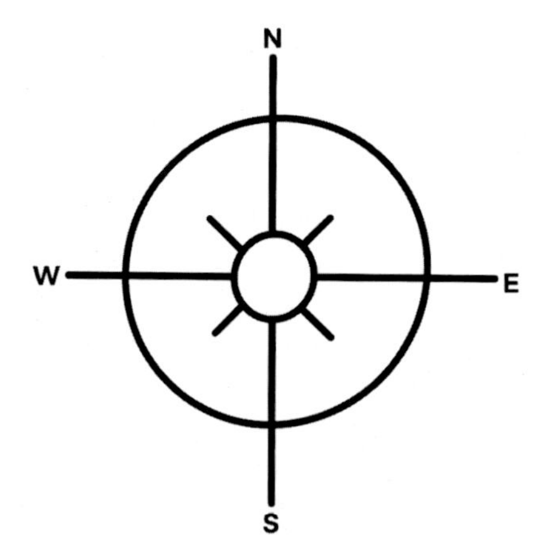

A symbolic representation of the 360° view from the Big Pinnacle (inner circle) to the horizon (outer circle) and beyond to the sky. This nearly perfect cardinal point aspect of Pilot Mountain was important to Native Americans for ceremonies throughout the year, tracking the rising and setting of star systems (e.g., Pleiades, Sirius, Orion), moon cycles, planting, and harvesting (courtesy of Lindsay Tillet McKelvey).

FOREWORDS

"I was spellbound by your eloquent words about Pilot Mountain. From the beginning of the book, you feel like you are reading about a mystery in nearly every page. It's hard to put the book down, even to gorge on a pork chop sandwich at Snappy's [Snappy Lunch] in nearby Mt. Airy. With my respect and admiration for your great work in making history come alive."

—RUFUS EDMISTEN
NORTH CAROLINA ATTORNEY GENERAL (1974–1984)
NORTH CAROLINA SECRETARY OF STATE (1989–1996)
MEMBER OF THE WATERGATE COMMITTEE WHO PERSONALLY SERVED PRESIDENT NIXON WITH A SUBPOENA. BORN IN BOONE, NORTH CAROLINA, AND LIVES IN RALEIGH, NORTH CAROLINA, WHERE HE PRACTICES LAW AND PARTICIPATES AS COHOST OF A WEEKLY GARDENING RADIO SHOW ON WPTF-FM.

"My expertise is in American figurative sculpture, but I have always been intrigued by the Easter Island statues and any large-scale statue or monument by ancient civilizations. In America, we have three sculptures that come to mind, carved from the sides and tops of mountains: Stone Mountain in Georgia, Mount Rushmore in South Dakota, and Crazy Horse in South Dakota, which is still underway. At times, especially earlier in each project, the blasting and carving of the stone of these man-made projects had the look of something made by primitive or ancient civilizations.

Then there's nature-made vs. man-made. I also offer information about rock formations in Bryce Canyon, Zion National Park, and Arches National Park in Utah. The 'hoodoos' in Bryce and Zion are amazing; some of them are quite lifelike structures. All of these structures are the result of wind erosion and especially water erosion brought about through freezing and thawing cycles over thousands of years. As I said at the beginning, I am not an expert on geological rock formations. But that may well be what you have at Pilot Mountain. Although, it is still nice to think that ancient Indigenous people had a hand in their creation, isn't it?"

—ROBIN SALMON
VP OF ART AND HISTORICAL COLLECTIONS/CURATOR OF SCULPTURE
NATIONAL SCULPTURE SOCIETY MEDAL OF HONOR AWARD WINNER IN 2022
BROOKGREEN GARDENS, PAWLEYS ISLAND, SOUTH CAROLINA

FIGURE F.1. This stone face does have a striking resemblance to an Easter Island Moai, but it is found at Pilot Mountain State Park. Is it just one of many natural coincidences on the mountain trails? When walking on the Ledge Spring Trail, just below the parking area, you must walk around this prominent feature. When looking straight at it, you can see a place on this stone face where it appears eyes used to be. The large block that looks like a top knot sits precariously on the stone face and has a notch to secure it to the mountain. There is an enigmatic tree growing out of the top of that block! Like many sculptures, it does appear to have been chipped away, but by humans or forces of nature? One wonders if this is just a stone face or if an entire body is buried in the steep trail. For the full impact of this truly mysterious stone face, it really needs to be seen in person. In this area, mountain climbers are allowed to scale the rock walls. Be warned, it is a strenuous hike to get to this location midway on the trail, and there have been falling accidents in this area where the Pilot Mountain Search and Rescue had to be called. Does this look natural to you?

ACKNOWLEDGMENTS

I would like to acknowledge the invaluable help received from the relatives of the former owners of Pilot Mountain: the granddaughter of W. L. Spoon, Frances Alexander Campbell, who put together over fourteen thousand items in the W. L. Spoon archives at the Wilson Library at the University of North Carolina at Chapel Hill (UNC-CH), and J. W. and Pearle Beasley's grandchildren, Betty Gay Shore (a.k.a. Betty Gay Shackleford, RN), Kenny Glace Jr., Beverly Beasley Glace, Dr. Charles "Gene" Glace, and Patricia "Patsy" Sperry, who all went way above and beyond. Their profound love of this mountain is now understood and that feeling shared by me.

I would also like to thank my friend Steven M. Weiss at the UNC-CH Southern Folklore Collection; the Wilson Special Collections Library at UNC-CH; Duke University's Rubenstein Library in Durham, NC; the Surry County Genealogy Association; Marion Venable and Annette Ayers at the Surry County Historical Society located in Dobson, NC; Pilot Mountain historian Carolyn Boyles; Amy Snyder at the Mount Airy Historical Museum in Mt. Airy, NC; Park Ranger Mark Farnsworth at Horne Creek Historical Farm; Sebrina Mabe at Surry Digital Heritage at Surry Community College; Fam Brownlee and Martha Rawls at Forsyth Public Library History Room in Winston-Salem, NC; and Adam Ressa at the Mocksville Masonic Lodge.

Thank you to Winifred Coleman and her daughter Deborah King for their help from Italy; J. Eric Elliott, archivist at the Moravian Archives; the State of North Carolina archives in Raleigh, NC; the Duke University history archives; Marcia Phillips at the Mocksville Public Library history room; the 26th Annual American Indian Heritage Celebration at the North Carolina Museum of Natural History in Raleigh, NC; Friends of Town Creek Indian Mound Inc.; State Library of North Carolina; the State Archives of North Carolina; George Crater;

and my friend and pilot for this book, Russell Joyce at the Mount Airy Regional Airport.

The help and encouragement are so appreciated from Dale Riddle of The King Bees; musician Neil Young for his "WOW" encouragement on his blog at NYA for this writing project on Pilot Mountain; Kevin "Thrasher" Booney for making me turn around to see the vernal equinox shadow of the mountain in 2022; Chloe Klingstedt at *Our State Magazine*; the extraordinary sacred site researcher and guide Emma Stow from the United Kingdom; and Dr. Robert Schoch and his wife Catherine "Katee" Ulissey.

Thank you also to Larry Melton and Matt Windsor, former Superintendents of Pilot Mountain State Park; current Park Superintendent Jason Anthony, who helped me the day the cover of the book was taken; Karen McAdams; Monika Muranyi; Barbara Marciniak; Dr. Al Goodyear of the South Carolina Institute of Archaeology and Anthropology at the University of South Carolina; musician Lucy Davis, who helped write "The Ballad of Pilot Mountain"; UNC archeologist Steve Davis; Director and State Archeologist John Mintz and Deputy State Archaeologist Lindsay Flood Ferrante at the NC Office of State Archaeology; *Winston-Salem Journal* writer Arlene Edwards Thompson; Jerry Venable; Jennifer Carroll; Mark O'Donnell; Bill Flinn; Paul Brennan; Tim Stone; Claudia Thomas; the National Archives and Records Administration; Katie Hall at the Raleigh office of the State Park Service; and Betsy Peters at Design Dimension Inc., who did the Pilot Mountain State Park new visitor displays and allowed a sneak peek at them before construction began. Cleve and Deborah Harris were extremely helpful, which displayed their deep appreciation of Pilot Mountain.

This book might seem a bit quote heavy, but that is because I wanted the ancestors who lived and learned before us to speak for themselves unfiltered about Pilot Mountain. Quoting the ancestors is a feature of the oral tradition of this land.

I would most especially like to thank my family for their understanding on the biggest writing assignment of my career, which took so many years to complete. I would also like to thank the wonderful people at MacFarland Publishing in Jefferson, NC, who guided this journey in the early stages of writing this book, and my two dear friends who helped so much with the professional editing of early drafts of this book, Nina Sazar O'Donnell and Lyneah Marks.

Thanks also to the staff at TIPS Publishing Services, including Elizabeth Coletti (project management), Claire Audilet (copy editing), Quinton Okoro (composition), and Coral Maxwell (indexing).

What a view even on a hazy afternoon on the Pilot Knob Trail of Pilot Mountain State Park just minutes off US Route 52 in Surry County.

Breaks in the clouds seen from the upper parking lot of Pilot Mountain State Park.

FIGURE I.1. Stone face on Pilot Mountain State Park trail.

INTRODUCTION

Have you experienced Pilot Mountain? If so, perhaps you will agree that there is something magnificent about this mountain that strikes a chord of awe in the human heart. For centuries, many have lauded the mountain's beauty with the oft-used term "natural curiosity."[1] This particular term describes Pilot Mountain State Park well even in modern times. There is a subtle allure during certain hours of the day when Pilot Mountain glows with a special luminosity dependent on the sun's daily and seasonal position in the sky. Pilot Mountain reflects the seasons so differently and accurately. From specific viewing points, this unique nationally recognized landmark in Surry County, NC, calls out for one to visit. That was exactly what happened to me one day as I drove by. It felt like an insistent doorbell waiting for me to finally answer.

According to legend, there are two footprints somewhere in the rock on the very top of Pilot Mountain—Noah's footprints, made as he stepped out of the ark. Pilot Mountain was actually Mount Ararat for a time, after the mountain in Turkey where tradition says Noah's ark landed. Travel a bit northwest of Pilot Mountain, and you will find a town named Ararat on the banks of a river of the same name. The name Mount Airy is also a derivative of Mount Ararat.

Pilot Mountain is a place of many legends—just as you'd expect. Anyone who has seen the distinctive mountain remembers it. It rises abruptly almost 1,500 feet from the surrounding Piedmont and is capped by a

1. Jeremiah Battle, "A Journey to the Pilot Mountain," *The North Carolina Star*, September 29, 1815: quoted in "The Pilot Mountain," *Tarboro Press*, June 29, 1844, 1. This long feature has been reprinted often with the last reprint found in the *Tarboro Press*, 2.

> large 200-foot-tall crystal knob. The knob, called Big Pinnacle, has light-colored, nearly vertical sides and is vegetated only on its flat top. Just west of the Big Pinnacle is a small saddle, beyond which is a smaller peak called the Little Pinnacle.
>
> One persistent myth is that Pilot Mountain is an extinct volcano, which, as we shall see, is not the case. Another legend has it that Indians, early settlers, and mountain men, including Daniel Boone, used the mountain as a landmark. There's no reason not to believe this one—such a visible and unusual feature no doubt guided many a traveler.
>
> —Kevin G. Stewart and Mary-Russell Roberson (2008)[2]

FIGURE I.2. Looking east at the Big Pinnacle of Pilot Mountain State Park, which was first photographed in 2012 with its many stone faces.

DISCERNMENT FOR THE STORIES OF PILOT MOUNTAIN STATE PARK

Discernment is needed when sorting through this collection of stories about Pilot Mountain. I do not claim to be an expert; many of these subjects were unfamiliar. So, I will be leaning heavily on the written legacy of experts in a variety of scientific fields in regard to Pilot Mountain. Sadly, written peer reviewed testimony from modern day experts about the mountain is hidden in redacted texts

2. Kevin G. Stewart and Mary-Russell Roberson, *Exploring the Geology of the Carolinas: A Field Guide to Favorite Places from Chimney Rock to Charleston* (Chapel Hill: UNC Press, 2008), 135.

with restricted access, which was found during research for these two books. Only the intriguing metadata was available from these archeological papers. We shall dive into that controversy in the second book, which reads more like an Ian Fleming spy novel than an attempt at sharing historical information about a famous ancient North Carolina landmark. These stories, collected from over the centuries in this first book, drive an interesting narrative that is well-documented with footnotes for further study. These are our elders speaking—during a time when newspapers had the stories that were then discussed on front porches across this land.

> Pilot Mountain is covered with legends. According to one legend, Indians used to send smoke signals from on top of the mountain. Early settlers sat about the hearth at night and watched the fires sparkle. They said somewhere on that mountain there was a footprint of Noah of Biblical times. He stepped off the Ark after 40 days and nights of rain, the worst soaking Surry County ever had. Once Pilot Mountain was believed to be a volcano.
>
> —Staff Writer (1978)[3]

Science says that Pilot Mountain is not a volcano, but many of these stories, legends, and lore are beyond the oftentimes narrow purview of modern science. Although, the recent discoveries in quantum entanglement might be catching up with ancient legends and lore from the oral history of this landscape. The detailed oral history of this area includes an invisible benevolent people who might be considered living in another dimension parallel to our own. Some will roll their eyes, but others will lean in a little more. Make no mistake, the purpose of this research was to seek the entire truth about Pilot Mountain. Determining whether this was a success or failure is up to the reader. During my on-the-job training, I was raised on the pillars of journalism: who, what, when, where, and why. These questions drove the research in directions that revealed surprising results. Lawyers will only ask questions in court when they know the complete answer before it is entered into evidence. The exact opposite approach was taken when researching Pilot Mountain State Park. After receiving some of the experts' short email responses, I had to get up and take a walk around the block to digest what was given, it was so shocking.

Maybe you haven't been on Pilot Mountain, but you have driven by or seen pictures and felt a spark of curiosity, even at a distance. Or perhaps the national news of the fire in 2021, which engulfed nearly all of the 1,066 acres of Pilot Mountain, captured your attention with concerns about people who lived nearby. If you have not yet visited Pilot Mountain, this book can open the door for you to explore this profoundly beautiful and mysterious mountain.

Pilot Mountain was called Stonehead Mountain in the early days. My initial introduction as a photographer was to the stone faces of Pilot Mountain, so the

---

3. Staff writer, "Pilot Knob—A Monadnock," *Mount Airy News*, July 27, 1978.

name did not surprise me when I discovered it in historic literature. This descriptive name confirmed there was a story to be found and told honestly. If you've been there, have you seen the subtly hidden stone faces? If not, the photographs throughout these two books will introduce you. Each chapter opens with a full-page image of the stone faces or stone animals seen along the Pilot Mountain State Park trails. If you have seen them, you might have wondered what they are or how they were made. Were they accidental or intentional? Why do they change so much depending on the time of day and year? Are they just weathered stone, man-made, or a bit of both?

These are some of the questions we will explore in this book. Is Pilot Mountain State Park the Mount Rushmore of an ancient civilization? Is its true meaning held deep in the shrouded, and often intentionally hidden, cobwebs of the history of mankind? Some of the enormous stone faces can only be seen from a distance of about a mile or more from the mountain.

What you see today is quite different from prehistoric times because everything was in a much larger context. The landscape included deep virgin forests of gigantic trees we can only dream of without paved roads, bridges, cell towers, power lines, or telephone poles. Massive herds of wild animals howled in the night and ran around during the day. It was not unusual for so many pigeons to fly overhead that it would darken the sky in the middle of the day. Bears, mountain lions, panthers, and all sorts of critters were found across the landscape of thick, tall, terminal, and seemingly endless forests.

That is just the recent historic past, but in the prehistoric past, a few thousand years ago, the megafauna included thirty-foot tall sloths, bears the size of a car, woolly mammoths, huge herds of woodland bison, saber-toothed tigers, mastodons, and so much more. Some of the birds in this megafauna time period had thirty-foot wingspans. This was right here in North Carolina, in the area of Pilot Mountain State Park, not some distant land across an ocean. Did the abundant food supply of the megafauna era also allow for giant eight- to fifteen-foot tall humans right in North Carolina? The oral history of prehistoric times and the newspaper history of historic times document giants. The how and why of the slow but steady die-off of the megafauna approximately thirteen thousand years ago is an ongoing debate. That the megafauna were here in North Carolina right along with humans of whatever size is now a fact. Can you imagine a thirty-foot sloth in your own backyard now—moving around chewing on the tree leaves? That certainly would get your full attention out the back window while you washed dishes, especially if a saber-toothed tiger attacked it while you were watching.

PASSING BY PILOT MOUNTAIN IN A CAR VERSUS HIKING ON IT

Like so many others, I had driven by Pilot Mountain State Park for decades but, while curious at that moment, had never stopped until one twilight in 2012. It is amazing that some people who live just a few miles from the mountain have never been on it, while others hike there with their dog every day. It was the light on the mountain that caught my attention driving south toward Winston-Salem, North Carolina on US Route 52 and drew me to stop.

"Light is the basic component from which all life originates, develops, heals, and evolves," according to Lady Carla Davis, ND, MPH.[4] Light is a funny thing; it can't be seen directly but makes all things visible.[5] Human eyes only see a tiny fraction of the full spectrum of light, so there are many invisible aspects to it before you right now. Twilight moments can make very memorable photos, and this early-evening bending of light from the horizon drew my attention, promising special photographs. Pilot Mountain State Park provided the perfect subject at the end of a long drive that day. After exiting off US Route 52, it was a pleasant surprise how quickly one reaches the mountaintop parking despite the multiple hairpin turns. The 20-mph speed limit up the mountain is no joke. Be aware that traffic goes both ways on blind curves. On the weekends especially, people hike right on the road rather than the trails, which is an added risk when driving up the mountain.

It was a clear, beautiful evening just starting to give way to twilight in 2012. There was time to take the short hike to the Little Pinnacle trail overlook for a look-see with a camera before closing. Once out of the car, the majestic vistas commanded attention from each of the viewing areas on the way to the overlook. It is a marvelous open-air view that is like being in an airplane but with your feet planted on the ground. While I didn't take the route, had the stairs still been there, one could make a good bet that I would have gone from the Little Pinnacle, across the saddle area, and then up the extra two hundred feet to the top of the Big Pinnacle. The photographs taken that twilight evening focused on the landscape and the sunset lighting. There was no rushed feeling during the quick visit, but the park ranger patiently waiting in his truck for the visitors to clear out might have wished there was. Seven cars remained, and his shift was not over until they all left. Later, I learned that one of the worst feelings for a park ranger is when someone doesn't return to their car after dark, which does happen. This is second only to a report over their communication devices of someone falling off a trail or precipice. The closing ranger has the responsibility to make sure everyone is off the mountain by the end of the day. Only then can the ranger close and lock the gate at the base of the mountain. On the day of the 2021 fire, the closing ranger must have been devastated to see thick smoke rising over the picnic area on a "no burn" weekend.

Once home after that first visit in 2012, it was a surprise to see stone faces in many of those images. They were small from that distance but stood out clearly in the evening light, which accentuated the details in shadows. The initial stone faces discovered were in photos taken of the Big Pinnacle's profile facing south and had gone unnoticed until the images were downloaded. I was intrigued and hooked, returning often to hike for miles and photograph other aspects of the mountain purposefully and not just as a scenic stop.

---

4. Lady Carla Davis, ND, MPH, "LIGHT – A Vital Nutrient," https://www.nourishingbasics.com/wp-content/uploads/2018/03/Light-a-Vital-Nutrient.pdf.
5. Harold W. Percival, *Thinking and Destiny*, 14th ed. (New York: The Word Foundation, 1974), 989.

By paying attention, the stone faces can be noticed in person, but more were always seen while editing photos, the best of which are shared in this book. The stone face artwork is so in tune with the natural aspects of the mountain that they blend into the surroundings. Even when telling the stories of Pilot Mountain, the meanings behind these faces in the stones remain mysterious. On one hand, science looks at them like an "it" or a bunch of rocks. The name Stonehead Mountain, to a scientist, is like naming a cloud that takes on a familiar shape. Indigenous people, however, look at Stonehead Mountain as a living "thou." They feel there is a consciousness imbued with sacred power that acts as a great guide. Listening, in stillness and quiet, to this very mountain is often rewarding for a human being. The state of North Carolina has issued a license plate now that has the initials, GG, that stand for great guide. Funds from the specialty license plate go to the nonprofit group Friends of the Sauratown Mountains that does good work there.

RESEARCHING PILOT MOUNTAIN ONLINE AND IN PERSON

There is a sublime balance of facts and mystery that surrounds the prehistoric era of Pilot Mountain State Park like an aura. Each return visit, I ran into hikers and rock climbers who had similar feelings as ravens flew over our heads. Curiosity sparked and grew with each visit. This led to many years of research all over North Carolina about the subject of Pilot Mountain. The research included dozens of difficult-to-read original deeds, tax records, books, newspaper articles, academic journals, paintings, woodcut blocks for newspapers, and early photography with detailed histories of Pilot Mountain. At first, the academic journals and thesis found in the research were freely available, but those specifically about the Pilot Mountain State Park landscape were restricted access or offered redacted. Isn't that very curious? That original spark of curiosity flamed, tracking down seemingly far-fetched personal accounts, like the tales of an airfield on top of the mountain, which caused serious doubts, even from the owner of the Mount Airy Airport. Ultimately, however, the stories turned out to be true.

> Jim, I have not heard of this either, but the Surry County Historical Society has an office here in the Historic Courthouse, where my sometime office is. Let me make some calls and see what I can get a lead on.
>
> On aerial photographs, I know some guys who would definitely help you out with that. I might be willing to help if you could get me a couple of pictures of my airport. We will make that work.
>
> I'll be in touch after I talk with the Historical Society.
>
> —George Crater (2020)[6]

6. George Crater, email correspondence with author, June 23, 2020. Mount Airy Airport is within minutes of Pilot Mountain by air.

Some outrageous stories of Zach Reynold's stunt plane flying also turned out to be true. There is even video evidence of his stunts in an independent movie by Joseph Wallace King called *Somebody Moved My Mountain*.[7] Doubt can be a real motivator for someone who likes research and digs ever deeper for answers. Redacted access to facts can be a real head scratcher, too.

Intriguing legends of a city beneath Pilot Mountain from the oral history of Native Americans reminded me of the stories of Telos within Mount Shasta in California, the Menehune of the Hawaiian Islands, and the ancestors of the Anangu around Ayers Rock in Australia. In addition, I interviewed people who shared their memories about Pilot Mountain. Many mentioned it was a shame I couldn't talk with a friend or grandparent who had passed away. They would often name a friend or relative who had trunk-loads of archives in their attic and many stories about the mountain, which were eventually lost to time. Despite this, many resources were found and shared, so these results follow with their permission.

RESEARCH SYNCHRONICITY

At twilight on the first day of the fire in 2021, I was taking photos from the Friendly Church parking lot that overlooks Pilot Mountain. Someone approached and asked me about the images I was taking. He suggested submitting them to the local paper, but when I said they were for a book I was putting together, his eyes lit up.

"You need to talk to some friends of mine who used to live right on Pilot Mountain," he said.

"Were they related to the groundskeeper, Bert Coleman?" I inquired without missing a beat, since he was one of the few people who had ever lived on Pilot Mountain who would still have living relatives.

"Yes, they are!" he replied emphatically with a twinkle in his eye and a warm smile.

Within minutes I was talking to the grandchildren of Bert Coleman, the groundskeeper during the transition from a private mountain to a state park over several decades. We spoke of the flagpole on the top of Pilot Mountain that their grandfather tended to in the morning and evening since there was no electricity. The flag had to be raised and lowered each day without the light on it. When lightning struck the flagpole, it wasn't replaced since it was such a chore, even though it was Bert Coleman's idea. We talked at length about the legendary King Bees, who played every Thursday night on the mountain just a couple hundred yards from the Coleman home. On any given Thursday night, the Dance Pavilion at Pilot Mountain was the most popular spot in all of Surry County. Growing up on the mountain and visiting grandparents near the stone house seen on the way up the mountain would have been interesting.

7. *Somebody Moved My Mountain*, directed, produced, written by Joseph Wallace King (Reynolds Films, 1975), https://www.youtube.com/watch?v=uB7hcolfcNY.

This entire first chapter wouldn't exist if I hadn't taken town historian, Carolyn Boyles, out to lunch in downtown Pilot Mountain. Sometime after that marvelous three-hour lunch, she attended a funeral and gave my contact information to Patricia Sperry after the service. Patricia Sperry is a grandchild of the last private owners, J. W. and Pearle Beasley. Everyone fondly called Sperry's grandmother Miss Pearle. Months went by before I got a phone call from Patricia Sperry after she found the misplaced note. She offered her help on the book and access to her grandmother's extensive archive of images and newspaper articles. Her input has been invaluable, and after consulting "dusty" newspapers and archives for years, her memories were a delight to hear. I now consider her a dear friend and am so grateful she introduced me to more of her family. Many of their stories are shared in this book and enrich it so.

Then there was the time my brother booked us a cabin at Pilot Knob Inn for the vernal equinox weekend in 2022. After booking it, I decided to go a bit early but only the cottage was available due to the spot's popularity. After booking the cottage, an appointment to interview Bert Coleman's sister-in-law, Winifred Coleman was made. She raised her family on the Pilot Mountain property, now the state park proper, and shared wonderful stories that I hope you too will enjoy. She asked where I was staying, and of all the places I could have booked, it was

FIGURE I.3. This image of the Pilot Mountain State Park wildfire was taken within twenty-three hours after it started on November 27, 2021. By that point, as the sun was setting, it was turning into a controlled burn of the entire mountain. At night, the orange reflections of the control burn flames made the smoke look like the fire was leaping high into the air. Damage to the trees was minimal, except on the Little and Big Pinnacle, which sustained extensive tree damage.

the two-bedroom cottage at Pilot Knob Inn, which she used to own and live with her beloved husband Ben and their children. During the autumnal equinox, we arranged to meet each other again for an entire morning. She showed me where the foundation of her house was on the mountain. We went up the road to the top of the mountain and back, which brought back her many memories of living there. We then went to her old home, now called the cottage, and she went through the house room by room. Doesn't everyone want to recall living in a place they used to call home? Afterwards, we sat on the original back deck that was built for her family and she told marvelous stories about her life.

I learned that Winifred Coleman's parents climbed the ladders with her two older siblings, but she didn't get to go to the top of the Big Pinnacle until there were stairs. Even then, with her husband Ben, coming down the stairs was the hard part. Ben knew so much about the mountain and he knew to climb down the steep stairs like a ladder. This way, you don' t have to look all the way down and can concentrate on the task at hand, getting down one stairstep at a time. She also remembered the tight entrance to the Devil's Den cave that opened to a large area within. The abundance of snakes around her home on the Pilot Mountain property was one of the main reasons she moved. The apple orchard drew rattlesnakes there during the Spoon and Beasley years of ownership on Pilot Mountain. Finding a black racer snake in her silverware drawer was the last straw for her. However, Ben tended about eight acres of tobacco on the Pilot Mountain property, so they didn't move far.

Some might call our meeting a coincidence, or a chance encounter, but I like to think of it as what Joseph Campbell defined as "following your bliss." When you are doing something you truly enjoy, the universe places things and people in your pathway to help you. Often this help is presented with a twinkle in the eye of the person helping, and this triggers your sixth sense. Isn't that the very definition of synchronicity? You still have to be in the right place at the right time. This often takes serious effort, but the reward makes it worth it.

WHERE ARE THE FOSSILS ON PILOT MOUNTAIN?

Did you know there are very few fossils on Pilot Mountain? Geologists say this is because the weathering process has eroded the mountain for millions of years. Pilot Mountain and some of the rivers around it date back at least 1.6 billion years. That is before there was anything alive and large enough to fossilize! Its weathered dome has guided travelers for years and it is now a guide for pilots. The dome, which looks like a petrified tree, is the revealed core of the original mountain. It's similar to the Devil's Tower in Wyoming, which was featured in the Stephen Spielberg movie, *Close Encounters of the Third Kind*. It is amazing how much Pilot Mountain State Park has in common with the Devil's Tower near Sundance, Wyoming, which we will address in the second book. Many think the rivers in this area around Pilot Mountain State Park are among the oldest, if not the oldest, rivers in the world. Given the 1.6-billion-year history of Pilot

FIGURES I.4 AND I.5. Just before sunrise on the day before the 2022 spring equinox, the arch of the setting of the moon, on the top looking south from Pilot Knob Inn, was near the Little Pinnacle. At the same location, the moon set on the other side of the Big Pinnacle the morning of the equinox, a half hour later on the bottom. Upon reflection, a video should had been taken of that breathtakingly beautiful setting of the moon behind the Big Pinnacle. Do you see the high point of the Big Pinnacle in these two images? That might be where the stone circle is located, where smoke signals by day and fire beacons at night originated in prehistoric times.

FIGURES I.6, I.7, and I.8. The construction of the new Pilot Mountain State Park Visitor Center is chronicled in these three images (more can be seen on the book's website). In the last century, this area (where the new Pilot Mountain Visitor Center is now located) used to be a very productive privately owned tobacco field. Part of the sales agreement for it to become Pilot Mountain State Park in 1968 included the harvesting of the tobacco after the sale to the state. The construction viewed above might be a modern update in a long history of earthwork construction at Pilot Mountain State Park.

Mountain, it has made many revolutions around the galactic center of the galaxy. Cycles within cycles within cycles seems to be a theme for life on earth, and Pilot Mountain has witnessed all of it. What happens every million years in space might be interesting to ponder, but remember these fascinating cycles exist on Earth too.

So many intriguing questions have remained unanswered and, until now, unexplored. This book is an attempt to open the doors and compile Pilot Mountain's down-to-earth past. In 1968, Pilot Mountain became a state park through the determined efforts of over twelve thousand people who put up hard-earned cash to buy it for the state and preserve it for future generations. For nearly two hundred years prior, Pilot Mountain was privately owned but even then it was open to the public with guided tours, campgrounds, swimming pools, and events from musical acts to bathing beauty contests that drew crowds. Before that, it was considered shared sacred ground by Native American tribes in this region who traveled hundreds of miles to visit it on foot. For recorded human history, life beyond thirteen thousand years ago is unknown. As more is discovered, it will be reported in this book series.

That awe towards Pilot Mountain is easy to feel but hard to describe. A hair-raising feeling, or goose bumps, pop up unexpectedly. It might happen on the expressway looking at it from afar, from the visitor center looking up at the almost perfect pyramidal shape of the mountain, or from the trails with the stone faces looking back at you. The images of the stone faces only hint at their grandeur seen in person.

The time it takes to get from the expressway to the top is amazingly short. The age of the road going up the mountain is a mystery, because it would seem the original road built in 1929 was just a rediscovery of an ancient one. Looking over the edge of the cliff faces 1,200 feet above the valley floor inspires awe. The view from the parking lot atop Pilot Mountain is two hundred feet higher than the top of the Empire State Building in New York City and with fewer obstructions to the horizon. Even at the parking lot, you still aren't at the top of Pilot Mountain. The view from the very top of the Big Pinnacle, where they used to have climbing ladders or stairs, is roughly 1,500 feet above the Piedmont floor and about 2,600 feet above sea level in all directions. The landscape of our planet rises and falls due to various pressures, so this is referring to the mountain's height in general terms.

Sometimes, while driving on US Route 52, the mountain looks like there's a tunnel ahead because it looms so large with the relatively flat Piedmont surrounding it. Pilot Mountain is distinctly oriented to the four directions, but from each of those directions it looks completely different. The Little and Big Pinnacle are formed on a nearly perfect east-west axis. Also, with the changing light, the view shifts hour by hour. On sunny days from a distance, the mountain takes on a blue hue, but not the green countryside or the Big Pinnacle.

There's a long history of people wanting their ashes scattered on Pilot Mountain. People in ancient and modern times must have felt something profound about this location to request it as a final resting place. Prehistoric legend has it that Pilot Mountain was not only the last resting place, but it was where relatives would come to pay their respects. Pilot Mountain State Park acknowledges in writing that there are prehistoric burials of "high significance" on the mountain but, beyond that, mum is the word.

This collection of stories about Pilot Mountain has taken me on an investigative journey, and I hope you enjoy this compilation of facts, observations, photos, tales, and legends as much as I did to track them down. It's important to know that there is controversy reported in the second book of this series that has wide-ranging significance to all human history. In this era of online access to research, without personally going to historical archives, nearly half this book would be overlooked or simply missing. Some of these archives were located hundreds of miles from Pilot Mountain, so even the locals who know all about Pilot Mountain might learn something new here.

FIGURE 1.1. Stone face on a Pilot Mountain State Park trail.

1

A Walk Through the Park

FIGURE 1.2. When you stand between the Big Pinnacle and the new Pilot Mountain Visitor Center, you are on the perfect east-west axis orientation of the mountain. In essence, that east-west axis is along the spine of the entire mountain that cuts right through the center of the pyramid that still retains its nearly pure crystal top.

First, we will get oriented with a bit of recent historical background on Pilot Mountain State Park in the first chapter. Many visitors just park at the top, walk to the Little Pinnacle, and are back on the road within an hour, which is fine if you are in a hurry. If you are one of the long-haul truckers who have driven by the mountain their entire working life, are retired now, and visiting Pilot Mountain State Park for the first time, this first chapter is dedicated to you. Hopefully this work will encourage a longer, more informed visit for everyone.

The second chapter will give a chronological timeline of the park's history starting in 1753. The second book will address the controversial, prehistoric mysteries of Pilot Mountain. Top experts of the state were contacted by email with questions, but their responses in writing created even more serious questions than answers. The methodical research for the second book went into some unforeseen directions that are documented in detail.

GUIDED TOUR OF PILOT MOUNTAIN

The day before the 2020 autumnal equinox was perfect for a guided hike on Pilot Mountain. The sky was, as they say, "Carolina blue," and clear of clouds. The ground was dry, but not dusty. There was low humidity for the time of year and a slight breeze, making it the perfect temperature. Everything was so green the mountain positively glowed from the base to the top. Two of the Beasley grandchildren, Beverly Glace and her brother, Kenny Glace Jr., had agreed to give a guided tour. They both grew up visiting the mountain and worked there during the summertime in their youth. They also attended the annual Beasley/Forkner reunions held there. It had been several years since they had been on Pilot Mountain, and they were excited to return and kindly help with the background for this book.

After this tour in 2020, I have made a point of visiting the mountain on the equinoxes and solstices to view both the sunrise and sunset. Like seeing the aurora borealis, you must be at the right place at the right time. Without all three factors (date, place, and clouds) in your favor, your viewing enjoyment can be in jeopardy on that particular visit. It should be noted that the perfect pyramid shadow is best on the equinox, but the pyramid shadow is seen year-round.

The Beasley family owned the mountain from 1944 to 1968 when it was then sold to become a state park. The firm stipulation of that initial sale agreement was that the mountain would remain in its primitive state, and if any development was done on Pilot Mountain, the deed would be returned to the family. Having been on the mountain looking for landmarks mentioned in various documents but unable to find them, I needed help. It was a joy for me to learn from living people, eye to eye, after digging through all the dusty archives for information. Archive request forms for librarians, boxes with folders, white gloves, parking passes, and folder place markers were all replaced with a tape recorder for interviewing. A Beasley granddaughter, Patricia Sperry, had tracked down her cousins Beverly and Kenny Jr. in Concord, near Charlotte, NC. She arranged for us to all meet in the spacious second-floor office of the Key City Antique Mall and Shops in North Wilkesboro, owned by Chris and Tammy Johnson. This interview was at a central location for all of us and would turn out to be our pre-hike meeting.[1]

Patricia, Beverly, and Kenny were young when their grandfather died in 1958. Thereafter, they helped their grandmother with running the mountain for the next ten years on and off until the mountain was sold. Beverly and Kenny now have children of their own. We talked at first by tag-team phone calls and agreed to meet at the antique store where Beverly ran a booth with Patricia. Patricia's mother, Carole B. Sperry (a.k.a. Ruth Carole Sperry), also spent a lot of time on the mountain as a child and later they both helped Pearle Beasley run the day-to-day operations in the summer. There was sort of an "aunties" coalition to help with the running of the mountain. It was expected for the thirteen grandchildren to help out on the mountain during the summer to assist their grandmother.[2] The Devil's Den cave was one of the places spoken of for a couple of centuries, and so was the pool area from the 1960s that had been elusive to find.

"Speaking about the caves and things like that, if you go around the Big Pinnacle at the base of it, there's several spots that you can go stick your head into," Beverly told me while sitting in an old antique chair during the pre-hike meeting at the office. "The last time I was up there, which was maybe about five years ago, we stuck our head in the cave that we always had. My mom would always let me play around the base, but not until I was nine was I ever to go up the steps. I'll never forget that family reunion where, 'Oh, I get to go up the stairs!' I was so excited."

"Because they were at an angle like this," Kenny continued the talk about the old stairs on the Big Pinnacle without missing a beat. His fingers were pointing almost straight up at a steep incline. "Let's put it this way: they would not pass any OSHA regulations."

We agreed to meet on the mountain on the last day of summer. During our initial encounter, they both had decided it was best to give an in-person tour of the mountain rather than write it down on a makeshift map. As we chatted about this future guided tour with our family and friends, the group of three turned out to be a party of eight. When it comes to Pilot Mountain, word spreads fast. One of the eight on our hike was Beverly's daughter, a great-grandchild of J. W. and Pearle Beasley.

We were all dressed for a hike. Kenny had a Tar Heels cap, a hiking stick, and a compass, which we consulted often. Beverly had on a blue cap with a Secretariat

1. Kenneth Glace, Beverly Glace, and Patricia Sperry (grandchildren of J. W. and Pearle Beasley), interview with author, September 1, 2020, North Wilkesboro, NC.
2. Dr. Charles "Gene" Glace (grandson of J. W. and Pearle Beasley), phone interview with author, March 2022.

logo. Our group now had four mature adults, all with a touch of gray hair, and four younger people, who were happy to climb around looking for cave entrances. We'd point in different directions: up, down, and sideways. We all met at the front of the beautiful, new Pilot Mountain Visitor Center at the base of the mountain. The temperature was in the low fifties and perfect. We piled into two cars and left the visitor center parking lot, which was about one quarter full.

"Welcome," Beverly greeted us the morning of the hike, her long strawberry-blond hair with a touch of grey flowing and her eyes dancing. She was happy to revisit her childhood memories and we were ready to listen.

"I just realized that we used to have free rein on Pilot Mountain as kids since we didn't have to pay to visit the mountain," Beverly told our group before we drove off with our guides in the lead car.

Beverly was on a mission with Kenny to help me find the old Devil's Den cave and the pool area that she remembered fondly from her youth. Having read so much about the Devil's Den on the Pilot Knob Trail (a.k.a., Jomeokee Trail) and Noah's footprint on top of the Big Pinnacle, our group's excitement to see it in person was palpable. Of course, the footprint in stone is now off limits, since the stairway has been taken down and climbing is forbidden. Some report more than one footprint in the stone on the Big Pinnacle, but there is no way to confirm that now.

This entire area used to be the land of the Cherokee tribe, covering several states, and Pilot Mountain was on the far eastern edge. Back then, this entire area of the Blue Ridge Mountains was not named but was referred to as "the land of the caves." In prehistoric times, these caves were used for many reasons, like mining. Copper from the Keweenaw in Michigan and mica from North Carolina were the two most prized possessions of the Native Americans. A unique form of mica is found right on Pilot Mountain, making it a valuable source in the prehistoric past.

> An integral part of the southeastern environment, the area holds evidence of prehistoric occupation from Paleoindian through Mississippian times. Eroded into the limestone underlying the region are thousands of caves (more than seventy-five hundred recorded in Tennessee alone), including many of the longest caverns in the world. Prehistoric peoples used these underground passages for at least the past four thousand years (cf. Watson 1974), and to consider only surface sites in the prehistory of the Southeast is to ignore a significant part of the region's archaeological record. In recent years, a growing number of caves that contain prehistoric art works have been identified, and it is now clear that cave art was produced in the region from Archaic to Historic periods. Much of the art may have religious or ceremonial relevance, and documenting this newly appreciated art tradition is becoming a major issue in a clearer

understanding of prehistoric southeastern, especially Appalachian, art and iconography.

—Jan F. Simek, Susan R. Frankenberg, and Charles H. Faulkner (2001)[3]

FIGURE 1.3. The Drink House used to be alongside the road going up the mountain, but it now sits above on the left going up the mountain. With the switch from private ownership to public ownership, the roads near the Drink House changed drastically. Some of the old roads have now crumbled and are part of the walking paths now.

THE DRINK HOUSE

On our way up the mountain, we went by the old, abandoned ranger station built in 1983, which is on the right side of the road after the circle turn going up the mountain. There is now parking there for walking on the trails, but the building isn't in use anymore. It would be a perfect location for a multidisciplinary archeo-astronomy team (i.e., archaeology, anthropology, astronomy, statistics, probability, and history) to have an office for research now.

Then, we went by the perfectly preserved old stone house that used to be called the Drink House, which had souvenirs and soda for sale. There is a driveway at the Drink House, just before you get to it, and parking to walk the trails. In that same area, there used to be the small Gate House, which was a fork in the

3. Jan Simek, Susan Frankenberg, and Charles Faulkner, *Archaeology of the Appalachian Highlands*, ed. Lynne Sullivan and Susan Prezzano (Knoxville: University of Tennessee Press, 2001), xxiv.

road where tickets were sold. The Gate House was a small room the right size for two people that was beside the road. It was locked each night when the park closed but has since been destroyed. Both buildings were made from the stone on Pilot Mountain. To the left at the fork, after you paid a dollar, you could drive up the mountain road to the top. To take the road to the right, you needed a paid pool pass, except on Thursday night, when an admission fee was charged for the dances. Beverly used to sell tickets, soda, and souvenirs as a teen there. The stone Drink House looked like it was built the day of our hike in 2020, but it is not in use right now. More recently, as the mountain transitioned from privately owned to a state park, the Drink House was used as a park office in the 1960s. The new Pilot Mountain State Park Visitor Center is now the third office for the park rangers since they took possession in 1968.

Further up the mountain, we went by the one of two mineral springs, named Boulder Cove, still flowing strongly right beside the road. Using these springs as landmarks isn't a good idea because there are so many in every direction, regardless of elevation. If you had grown up here, the major springs would be easy to locate. The Native American water keepers, who are often grandmothers, would be well aware of these landmarks. They are very active around the Great Lakes region, drawing attention to water issues, but are now also found across the planet. The major mineral springs that eventually flow to the ocean are noted on maps for Pilot Mountain State Park. This waterflow from the mountain is one of the reasons to not use chemical spray paint on Pilot Mountain stone, as it pollutes the drinking water at its source.

As we drove by Boulder Cove, Beverly and Kenny pointed to it.[4] That little spot was used in the past for overheating cars. I could imagine the men in their Sunday best getting out of their cars here to fill their steaming radiators. That cool, refreshing, mineral spring water is in a beautiful, deep green, dense forest with a mixture of coniferous and mostly deciduous trees (e.g., Chestnut Oak, Black Pine). Clearly, the mineral spring keeps the trees and undergrowth happy. The ravine there, one of so many on the mountain, might have been cut by the water flowing down when there was much more snow on top of Pilot Mountain. A form of roadside assistance back then, it was now abandoned. About a month after the hike, another cousin, Betty Gay Shore, brought up the ladle for that mineral spring during a phone interview.

"Keeping that ladle to dip into the mineral spring water was often an issue as people would walk off with it after they tended their steaming radiators," she reported. It is no longer really visible from the current road. We could easily have missed it if we were not with our guides who pointed it out.[5]

4. Harvey Dinkins, "Dream Comes True in Opening Road to Top Pilot Mountain," *Winston-Salem Journal Sentinel*, July 14, 1929.
5. Betty Gay Shore (eldest granddaughter of J. W. & Pearle Beasley), phone interview with author, September 23, 2020.

FIGURE 1.4. This sign is a very important warning when driving on the Pilot Mountain State Park Road.

PILOT MOUNTAIN HILL CLIMB RACE

"Somebody might have approached him at the dealership or something," Kenny recalled. Kenny would have been about ten years old when the Pilot Mountain Hill Climb started. "He had the Ford dealership and the Chevrolet dealership."[6]

Driving up the road, some in our group recalled the early 1950s, when J. W. Beasley closed the mountain for the sports car races using the same road. J. W. Beasley allowed the Sports Car Club of America, founded in 1944, to have an amateur sports car race up the mountain starting in 1952.[7] It was called the Pilot Mountain Hill Climb and became a tradition that continued for a number of years. The Sports Car Club of America doesn't have records going back that far, so it isn't clear why they chose Pilot Mountain.

In order for a sports car owner to get a competition license he must first pass a physical examination. Then he usually attends a driver's school for several months. He must serve some time as a novice before he becomes a regular driver, [Julian] Putney said. After he has participated in three

6. Glace, Glace, and Sperry, interview.
7. William Blizzard, "Tom Thumb Tearabouts," *Sunday Gazette Mail*, August 9, 1959. Now the *Charleston Gazette-Mail*.

> events and has satisfactory reports he can get a competition license making him eligible to participate in national and regional events.
>
> —Dorothy Brimer (1959)[8]

Both Beverly and Kenny remembered those races fondly but didn't see how they could be repeated now in our litigious society. Individual sports cars like Porches and Ferraris were timed to see who could have the shortest time in this nationally sanctioned event. The Sports Car Club of America still exists all these years later. The sports-car traffic was one way up, using both lanes to reach the top of the mountain. Spectators, who paid to watch, lined the sides of the road. Even though their grandfather allowed the races, Beverly and Kenny didn't know how the races were timed. The one with the fastest time won the award for their division.

On August 14, 1953, the Statesville Daily Record reported that the Sports Car Club of America had four thousand members. Only one news item about a serious injury could be found about the Pilot Mountain Hill Climb race: A driver from Burlington, NC, received a head injury during a trial run of the race. His head injury resulted in fourteen stitches when his MG sports car turned over on a sharp curve on Pilot Mountain.[9]

"I remember I saw one race," Beverly stated during our pre-hike talk. "The cars would go up the newly paved road one car at a time using both lanes to see who could get the fastest time up Pilot Mountain. They would close the mountain while the race was on and you could buy a ticket to be a spectator."[10]

"Especially once you get at the very top it goes whooooo," Kenny Jr. exclaimed. "But some of those turns, you know, you're just, wow! That was before the days of attorneys."

The last few documented races in North Carolina were in 1957 at Chimney Rock in an event that drew sports car drivers from five states. There was also the Grandfather Mountain Hill Climb in 1960, which was the last of races there because of the fear that spectators would be injured.[11] However, the Grandfather Mountain Hill Climb was later revived in 1985 with Porsches, Ford Formulas, Saabs, and Corvettes in the race. Over one hundred signed up in the seven categories and twenty-four classes (based on size of the engine) of sports cars.[12]

"I have heard about the race thing just recently; they tried to do it again," Kenny recalled. "They wanted to rent Pilot Mountain, either for the day or for the

---

8. Dorothy Brimer, "Julian Putney Likes High Speed—On Ground in Racers, in Sky in Planes," *Kingsport Times News*, June 15, 1958, 18.
9. Staff writer, "Malpass Hurt in Accident," *Burlington Daily Times News*, August 30, 1954, 12.
10. Glace, Glace, and Sperry, interview.
11. Staff writer, "Ben Warren Named to Local Post by Sports Car Club of America," *Burlington Daily Times News*, October 17, 1960, 17.
12. Staff writer, "Grandfather Mtn. hill climb set this weekend," *Burlington Times-News*, June 13, 1985, 3b.

weekend. The state just said, 'Oh, no way.' Then there was a huge outcry that they shouldn't be allowed to have a private thing on the mountain like that."[13]

These days, one can race in the Pilot Mountain to Hanging Rock Ultra on foot. There are three races: the 50 Miler, the 50K, and the 50 Miler Relay, but competitors must register early because these events fill up quickly.

FIGURE 1.5. Harrison's Signal Rock is on the right side, just before the final hairpin turn towards the upper parking lot.

HARRISON'S SIGNAL ROCK

Off to the right, near the top of the road, is a huge rock known as Harrison's Signal Rock. This is before the last hairpin turn into the parking lot at the top of Pilot Mountain. This rock was found in 1929 while clearing the area to build W. L. Spoon's clay and sand road. It looks like an arrowhead for a giant. They named the rock after the steam shovel operator who found it, Mr. Harrison. It has stood in the same spot for over nine decades now.[14]

J. W. BEASLEY OVERLOOK

As we made the hairpin turn into the parking lot, we passed the spot where J. W. Beasley liked to have people pause in their cars. In our pre-hike talk in North Wilkesboro, Kenny said this was his favorite spot on the entire mountain because of the incredible view. In those days, the road was much closer to the cliff with a secure rock wall on the curve that was built by Pearle Beasley. Later, Betty Gay

13. Glace, Glace, and Sperry, interview.

14. W. L. Spoon, "The Story and Facts about the Pilot Mountain in Surry County, North Carolina," box 10, folder 436, 8, W. L. Spoon Archives, Wilson Library, University of North Carolina-Chapel Hill.

Shore said the view from this spot that her grandfather loved would "take your breath away" with the clear view of the Sauratown Mountains on the left and the Big Pinnacle on the right.[15] He used to keep trees cut back so the view of the Big Pinnacle was perfectly clear. It would be wonderful if the Beasley Overlook view was revisited today, keeping the trees trimmed. It truly is a special historic spot on the mountain, even though seventy-five years later the trees block the view. There is a short trail called the Sassafras Trail below that hairpin turn, which also offers a spectacular view. This is a different orientation to the Little Pinnacle overlook view of the Big Pinnacle and the Sauratown Mountains, but slightly lower in altitude from the Beasley Overlook, and it's highly recommended.

GEOLOGICAL HISTORY OF PILOT MOUNTAIN

There is a marvelous pamphlet that the Pilot Mountain State Park made available in 1994, which readers can print online from a PDF. Readers with children might appreciate that this goes into great detail about the unique geological history of Pilot Mountain. There are experiments parents can do at home and it is a great guide once you are at the park. According to the pamphlet:

> One billion years ago, this area was a shallow sea. For hundreds of millions of years, layer upon layer of sediment accumulated on the sea floor. The pressure of the overlying layers of sediment caused the formation of sedimentary rocks, such as sandstone. During this period, there was also some volcanic activity in the area. Approximately 700 million years ago, magma came up through the cracks in the layers of sediments and eventually cooled to produce granite, an igneous rock. Igneous rocks are formed from the cooling of lava or magma in volcanic activities. Lava is magma that reaches the surface of the earth. The type of rock formed depends on the rate of cooling and the mineral content. . . Quartzite, the rock formed from metamorphosed sandstone, is much more resistant to weathering than the other types of rock in the area. It acted as a cap over the underlying rocks, protecting them as the surrounding landscape was worn away. The resulting pinnacle formation is known as a monadnock. Pilot Mountain is a spectacular example of this phenomenon; many monadnocks are simply ridges or rocky hills.
>
> Today, you can witness one billion years of history at Pilot Mountain State Park.
>
> —Michael Smith (1994)[16]

15. Shore, phone interview.
16. Michael W. Smith, *Jomeokee Geology: An Environmental Education Learning Experience (Pilot Mountain State Park, 1994)*, 13–14, https://files.eric.ed.gov/fulltext/ED376053.pdf.

PARKING LOT ON TOP OF PILOT MOUNTAIN

Atop Pilot Mountain, near the Little Pinnacle, visitors used to be able to park anywhere on the unpaved lawn that was pounded to hard, beige dust in the hot, dry summers. There were shade trees all around the perimeter of the parking area. These days, there are reserved handicapped spaces between the white lines on a paved surface. During the Spoon and Beasley ownership, nearly five hundred cars could park on the top. Now, paved parking only allows a fraction of that number, even with the upper and lower parking lots combined.

In August 2020, a three-year trial project began with a twenty-two-seat van with wheelchair space to take people to the top. This is for when that parking lot is full and it also saves visitors' vehicles from wear and tear. During peak times, the Pilot Mountain State Park estimates that over three hundred cars drive up the road to the summit each hour. The shuttle runs on weekends and holidays from March to November for a small fee.[17] This seems to have resulted in a lot less traffic up the mountain and much more hiking.

AIRFIELD ON TOP OF PILOT MOUNTAIN

Where the picnic area is now there used to be a large parking lot, though that wasn't its initial purpose. The original plan was for an airfield for early airplanes and gliders. It was located where picnic tables are now surrounded by trees, near

FIGURE 1.6. A B-52 bomber practicing low-level flying to avoid radar detection in 1965. Photo courtesy of Frank Jones Collection of the NC Room at the Forsyth Public Library in Winston-Salem.

17. "Shuttle Service," FAQ, Pilot Mountain State Park, accessed November, 21, 2022, https://www.ncparks.gov/blog/2022/09/25/summit-area-closed-vehicle-traffic-weekends-and-holidays.

the Little Pinnacle, but there is no sign of it now. The airfield was downhill from the current paved parking area going west and was said to be 1,400 feet long. The picnic area used to be at the top of the airfield much closer to the Little Pinnacle, where the wooden fences are now. After the park service took over, the picnic area has been moved and is on the downhill slope of the airfield. The picnic area was much easier for the elderly to access before than the way it is now. Nature has reclaimed the airfield to the point that unknowing viewers wouldn't know it was ever there.

One pilot, who wanted to remain anonymous, said it was the only airfield he'd ever landed on that he had to use full throttle for the entire landing since it was so steep. To take off again, he had to get to the very top of the airfield to have room to clear the trees at the bottom! During the chapters on the two private owners of the mountain, W. L. Spoon and J. W. Beasley, we will explore the airfield in more detail. It is interesting to note that the variable 22.1 and 24.5-degree tilt of the earth is similar to the tilt of the former airfield. This connects to the questions about aspects of Pilot Mountain State Park being natural or man-made.

MILITARY USE OF PILOT MOUNTAIN

Moving toward the Little Pinnacle on the paved sidewalk, Beverly pointed out the stone walls, the construction of which was supervised by Pearle Beasley. This felt like a grand tour of Pilot Mountain State Park, and it only got better. The Beasley grandchildren had been here many times and were proving to be knowledgeable guides. They walked and guided us with such intention, while still pointing at things of interest along the way. There are several areas to walk out to an overlook. Those who weren't keen on heights sat on the stone wall away from the incredible overlook 1,300 feet or so above the Yadkin Valley. While clearly not for the faint-hearted, the view alone from the parking lot is a draw to Pilot Mountain.

On our autumnal equinox hike with the Beasley grandchildren, a huge circling military helicopter made close passes around the Big Pinnacle, which has been a regular occurrence for decades. It was so close that it was difficult to tell if there were one or two helicopters. Kenny had been a lifeguard at the pool on Pilot Mountain back in the 1960s. He said the helicopter reminded him of the Boeing B-52 Stratofortress bombers, which practiced flying low to avoid radar detection while he was working.

"The B-52 bombers were much louder. The helicopter noise did shake the air, but the B-52 bombers would shake the ground," Kenny remembered. "You could set your watch to the timing of the B-52 runs over Pilot Mountain," which occurred every day at 3:00 p.m. "It felt like they were flying at treetop level."

> Once used by Indians as a landmark, the mountain is now used by huge B-52 jet bombers as a guide for daytime navigation. Flying to defend the skies of the United States takes practice, and that's what these big jets

that fly over the Sauratown Mountains and Tanglewood Park are doing. These B52 heavy bombers are on routine navigational missions. They come over this area from the southwest, fly at about 3,000 feet over the Sauratowns then disappear into the west. These flights are just one of possibly 20 over the United States that planes of this type take to maintain navigational proficiency. The big planes are powered by eight jet engines and carry a crew of six. These missions are flown only in good weather and away from commercial airlines' flight patterns. Planes of the Strategic Air Command are in the air 24 hours a day somewhere over the United States and can be sent instantly to almost any spot in the world.

—Staff writer caption to Frank Jones image above
for Winston-Salem Journal (1965)[18]

FIGURES 1.7 and 1.8. What a view from the Little Pinnacle on a windy day for the hats of Pilot Mountain Hawk Watch volunteers in September of 2020 (above) and 2022 (to the right). A Cherokee clan member could have some valuable insight about migration week from their oral history.

Air Force rules require planes to be three wingspans above the land surface. For a B-52, that would be a little over eight hundred feet—certainly close enough to wave at the pilot from Pilot Mountain. Officially reported in the Winston-Salem Journal public relations statement by Frank Jones, the planes flew at three

18. Frank Jones, image of B-52 over Pilot Mountain with caption, *Winston-Salem Journal,* July 19, 1965.

thousand feet. A librarian in the town of Pilot Mountain mentions that huge transport planes in 2021 make the run the B-52 bombers used to make. They rattle his dishes on a regular basis but it is just something that "you live with."[19]

PILOT MOUNTAIN HAWK WATCH

"Look at that!" someone shouted far ahead of us on the path to the Little Pinnacle.

It isn't just the military that flies over Pilot Mountain regularly. Huge raptors display their wide wingspan where the Hawk Watch volunteers station themselves. On the day of the tour, the loud military aircraft was disturbing even the local hawks with all the noise; it was mesmerizing to watch. There are certain days when it's difficult for planes to fly over the mountain, and the day of our hike happened to be one of them. Those old enough to remember the bird strike that brought Captain Chesley "Sully" Sullenberger's plane down on the Hudson River will understand this. Pilot Mountain is located in the midst of a heavily used bird migration path going south for the winter. There is a wonderfully detailed display explaining this migration path in the Pilot Mountain Visitor Center.

The Pilot Mountain Hawk Watch count didn't start until October 4, 1973, because until then, nobody noticed it was a major migratory path for raptors. Thanks to the Audubon Society of Forsyth County chapter founder, Ramona Snavely, who discovered the connection between wind patterns and the birds' journey through her research, Pilot Mountain Hawk Week began in 1973. The Little Pinnacle at Pilot Mountain was chosen as the location to count them and still is all these years later.[20]

Our hike began by walking up the stone steps to the Little Pinnacle. There is a huge all-weather poster on that path depicting the life-sized wingspans of the birds being studied. At the Little Pinnacle overlook, we found that it was staffed by three Hawk Watch volunteers. There was also a friendly little hawk on a rock just outside the fencing, so close that you could imagine petting the top of its head. The Little Pinnacle observation area is one of two hundred sites on the Migration Association of North America's watch list now. The volunteers were already set up and working when we arrived from our hike. They displayed a whiteboard listing the different types of birds, which they updated. They each held binoculars and had dressed in layers for the changes in temperature. The volunteers all looked south, where the thermals form over the exposed rock.

A thermal is rising hot air. It is interesting to fly in a small airplane around Pilot Mountain. Even on a calm day, you can feel the hot air forming thermals as soon as you cross the mountain from the north to the south. Those thermals can kick up, causing a bumpy ride, even when it has been perfectly calm during the short journey from the Mount Airy Airport to the mountain. Should you fly around the mountain, be aware of that change with the thermals if you are

19. Pilot Mountain Library employee, phone interview with author, 2020.
20. "Pilot Mountain Hawk Watch 2021," Activities, Forsyth Audubon, accessed November 28, 2022, forsythaudubon.org/hawk-watch/.

FIGURE 1.9. Many birds are seen flying around the Big Pinnacle in contrast to the clouds.

holding a cup of coffee or a camera. The heat rising off the rocks on the sunny, south side of the mountain intensifies the thermals, which both the local and migratory birds enjoy riding with ease. Photographers have observed that the thermals' heat is so intense that it bends the light, blurring images taken, especially of the south side of Pilot Mountain from any distance.

The first observation of migrating hawks at Pilot Mountain State Park was made on 4 October, 1973. In the ensuing years numerous sightings have been made of raptor species during the autumnal migration. The subsequent sighting and unusual abundance for a relatively unknown fly-way prompted more detailed study and observations. As is generally found at most look-out stations, the Broad-winged Hawk demonstrated the most dramatic migratory flights, often appearing in thermals containing several hundred circling hawks. Additional sightings of Bald and Golden Eagles provided the flyway with heretofore unreported species. Following a cold-front passage, the strong northwest winds produce updrafts as they strike the mountain range. Under such favorable flight conditions, the greatest number of individuals was observed in September. This same weather pattern during October and November played no significant role in migratory patterns. The northeast wind following a high-pressure system located northwest of the site, however, created conditions for the largest variety of species for each of the three months. Much is being learned about migrating raptors and the factors influencing their movement as more research is undertaken. However, there still

> remain many unsolved questions concerning the essential ingredients which cause hawks to move when known influences are absent.
>
> —Ramona R. Snavely (1973)[21]

The best viewing day at Pilot Mountain happened in 1993 when eleven thousand broad-winged hawks flew by on the thermals. Bald eagles, peregrine falcons, and osprey are also observed flying south, however a majority of all raptors counted are broad-winged hawks. In the springtime, they migrate over a different landscape when they return north. That was an exceptional day in 1993, but normally a total between two thousand and five thousand broad-winged hawks were witnessed flying south each season for the count. One wonders how many eagles and hawks gathered here in ancient times and provided feathers for Native American headdresses. The informative displays set up along the trails to the Little Pinnacle create an educational experience for young and old alike.[22]

WHEN PIGS FLY

Bald eagles sometimes fly around Pilot Mountain during their migration. The following explanation of bald eagles in North Carolina is from John Lawson's observations of the various birds he found here in North Carolina over three hundred years ago. What a story it is! According to most birders, the bald eagle only has one nesting time of the year and they lay usually two, but sometimes three, eggs. If the nest is disturbed, they might lay a second batch, but that is rare. Lawson's account from 1701 differs from the current research saying they have one nest per year. Bald eagles faced near extinction because of a pesticide, DDT, so one wonders if that affected fertility and the number of eggs they lay per year. In his reports, John Lawson references the common idiom: "when pigs fly."

> As the eagle is reckoned the king of birds, I have begun with him. The first I shall speak of, is the bald eagle; so, called, because his head to the middle of his neck and his tail, is as white as snow. These birds continually breed the year round; for when the young eagles are just downed, with a sort of white woolly feathers, the hen eagle lays again, which eggs are hatched by the warmth of the young ones in the nest, so that the flight of one brood makes room for the next, that are but just hatched. They

21. Ramona R. Snavely, "The Influence of Weather Conditions on Hawk Migration at Pilot Mountain State Park, NC," Forsyth Audubon Society, 1973, http://forsythaudubon.org/wp-content/uploads/2019/08/Ramona-Snavely.InfluenceofWeatheratPIMO.HW_.pdf.
22. Jane Borden, "Middle Mountains: Pilot Mountain," *Our State Magazine*, February 24, 2020.

prey on any living thing they can catch. They are heavy of flight and cannot get their food by swiftness. To help them there is a fish hawk that catches fishes and suffers the eagle to take them from her, although she is long winged and a swift flyer, and can make far better way in her flight than the eagle can. The bald eagle attends the gunners in winter, with all the obsequiousness imaginable, and when he shoots and kills any fowl, the eagle surely comes in for his bird; and besides those that are wounded and escape the fowler, fall to the eagle's share. He is an excellent artist at stealing young pigs, which prey he carries alive to his nest, at which time the poor pig makes such a noise overhead, that strangers that have heard them cry, and not seen the bird and his prey, have thought there were flying sows and pigs in that country. The eagle's nest is made of twigs, sticks and rubbish. It is big enough to fill a handsome cart body, and commonly so full of nasty bones and carcasses that it stinks most

FIGURE 1.10. Two of our Pilot Mountain State Park cave explorers, who were a fantastic help on this last day of summer, on the Little Pinnacle overlook with the Big Pinnacle as bokeh for the image.

offensively. This eagle is not bald till he is one or two years old.

—John Lawson (1709)[23]

23. John Lawson, *A New Voyage to Carolina: Containing the Exact Description and Natural History of That Country* (London: s.n. 1709), 137.

After chatting with the Hawk Watch volunteers, enjoying the Little Pinnacle view, and taking pictures, we headed on down the stone steps toward the parking lot and turned right. Walking a short way then taking another right, we continued on towards the Big Pinnacle on the .8-mile Pilot Knob Trail. On the way, we paused to see what looked like a stone snake coming out of a stone cave down to the right, just as we left the fenced-in part of the Little Pinnacle viewing area. This area might be the most popular spot on the mountain, but there was so much more to discover.

Next, we headed to the Big Pinnacle, using the trail found just around the corner from the bathroom and drinking fountain. This is a wonderful trail often missed by visitors who just go up to the Little Pinnacle for the view and then leave. The Pilot Knob Trail all the way around the Big Pinnacle can be hiked in about an hour, but photographers like to linger. There is a stunning image to be found in almost every direction on the Pilot Knob Trail. There is an old photographers' saying, "if you see something in front of you worthy of an image, look behind you since it might be just as amazing." We all navigated the stone steps down the trail and then we came to the Ledge Spring Trail turnoff. When you come to that fork in the trail between the two pinnacles, you have to decide which way to go. This area in the saddle has eroded differently and faster than its surroundings, which can also be said of the Piedmont in relation to the mountain.

> At one point, this whole region was a peneplain, or flat area. The Big and Little pinnacles are the result of millions of years of mechanical and chemical weathering. Weathering is the breaking down of rock by action of water, ice, plants and chemicals. Weathering wears down all rocks but at different rates. Examples of mechanical weathering are: ice wedging, water seeping into cracks in the rock and freezing, the expansion causing the rock to split; and root destructions, plants growing in crevices and cracks which apply pressure to the rocks causing the cracks to widen. Chemical weathering occurs when some minerals in the rock react with the air and water, dissolving the materials holding rocks together.
>
> —Michael Smith (1994)[24]

After making the short trip to the center of the saddle area, we paused for a group huddle to decide which trail to take next. There were three options: going east on the Pilot Knob Trail clockwise, counterclockwise all the way around the Big Pinnacle, or west on the Ledge Spring Trail, which goes below the parking lot. All three options end up circling back to the parking area. After a bit of discussion, we decided to continue clockwise around the Big Pinnacle on the Pilot Knob Trail. That way is less strenuous for the elders and starts in the cooler shadow of the mountain. Both trails feature impressive stone faces, but the Ledge Spring

24. Smith, 58.

Trail is much longer and more strenuous. There is a sign at the start of the trail that says it is like hiking up a ten-story building ten times. That statement is not far off the mark. The Ledge Spring Trail is where the mountain climbers practice.

We all followed Beverly closely because she pointed out many stone faces not seen before. What Jimi Hendrix is to guitar, the equally impressive Beverly is to finding stone faces. We were all in awe. As we walked around the Big Pinnacle, a crowd gathered to listen to the stories of our two guides. Overlooking the new Pilot Mountain Visitor Center halfway round the Pilot Knob Trail, the stone face and body of what might be the largest stone figure at the mountain was pointed out by the group to our two guides. We were proud to be able to give back a little to those who had given us so much on this hike.

ONWARD TO FIND THE DEVIL'S DEN

The Devil's Den cave area, the name of which may refer to its blackened overhang, was mentioned repeatedly in newspaper articles about Pilot Mountain over the last two centuries. There is another location in Madison County further west with the same name.[25] W. L. Spoon makes note of it in the 1929 pamphlet that he sold for ten cents at the road opening.

> The wonderful thing about this little cavern is the fact that at all times a breeze blows out if it strong enough to put out a match. It's cool. That is a misnomer. The old boy has moved out.
>
> —W. L. Spoon (1929)[26]

> Among the mysteries of this region is the 'Devil's Den,' a small grotto from which nearly a steady breeze blows at all times, strong enough to snuff out the flame of a match. In this strange hole is enough air-conditioned floor space to accommodate a party of overnight campers, but this of course would require the permission of the owners."
>
> —R. S. White, Jr., (1960)[27]

While the Devil's Den was mentioned often in the past, there was never mention of any prehistoric rock art found in the caves. As we continued our hike, we looked for signs of the cave. It wasn't as easy to find as we had hoped. The Devil's Den area is also featured in the 1994 geology pamphlet:

---

25. "Devil's Den and Noah's Footprint," *The Western Sentinel*, July 14, 1922.
26. Spoon, "Story and Facts about the Pilot Mountain," 8.
27. R. S. White Jr., "Jomeokee: The Great Guide," *Ford Times*, November 1960, 23–25.

> Devil's Den was carved by the force of moving water. One of the most interesting features of this stop is the small cave that is just above and to the left of the Devil's Den. Climb up to feel the wind coming out of the cave. The cave extends all the way to the top of the Big Pinnacle. The air movement is caused by the difference in air temperatures inside the cave. As the cool air sinks more air in pulled into the top of the cave. This air also cools and falls once it is inside the cave and shielded from the sun. In the winter, the air inside the cave is warmer than the air outside.
>
> —Michael Smith (1994)[28]

This type of weathering is the basis for the formation of caves, and the nearby Linville Caverns is another example of this. The full extent of the cave system within Pilot Mountain is unknown, though Beverly mentioned hearing the sound of moving water when she was young. Those who used to crawl in the tiny space of that cave said it opened up into a large room and went down, as Winifred Coleman confirmed. According to several independent sources, when you dropped a stone within the cave, you couldn't hear it hit bottom. We were all looking forward to experiencing the change in temperature and wind that Beverly mentioned. We also hoped to hear the running water within the mountain. Is this cave entrance the start of something as grand as Mammoth Cave National Park in Kentucky?

> No matter how hot it was outside you could stick your head in that little narrow cave entrance and the cool breeze from inside the mountain would blow your hair back. It was like standing in front of an air conditioner on a hot day in the summertime. If you listened very closely you could hear water running inside the mountain at the same time.
>
> —Beverly Glace (2020)[29]

While looking for the cave at the Devil's Den, we came across a warm spot. Our young friends from Durham called us back on the trail to feel this unexpected shaded warm spot. We were on the sunny south side of the Big Pinnacle, but it was odd that this shaded area was markedly warmer than in the sunshine. Without a cloud in the sky, the UV rays were pretty strong. We wondered if this was an area being warmed by a vent from inside the mountain or a cave entrance we did not find, because there weren't rocks in that area to radiate heat. All eight of us looked around but couldn't find the source.

28. Smith, 58.
29. Glace, Glace, and Sperry, interview.

FIGURES 1.11 and 1.12. We attempted to enter the Devil's Den cave the day of our hike on the equinox eve of 2020 (left). Months later, a Beasley great-grandchild on a different guided tour attempted to get in the Devil's Den cave unsuccessfully (right). Clearly, it is a tight fit to even get that far into the cave now. It is humbling to go into the cave since you have to go in on all fours. This humbling aspect to this cave would have been important to Native Americans.

What might be on the walls of the Devil's Den cave? Behind the debris that now blocks the entrance to the Devil's Den, could there be the rock art of "high significance" noted by Pilot Mountain State Park? In light of this new plasma research and that connection with petroglyphs, Pilot Mounain's rock art would be a fascinating discovery.

More than one hundred years ago, researchers demonstrated that ancient myths and pagan religions are similar throughout the world. Other researchers have known for years that ancient petroglyph images are similar throughout the world. What was not known until near the end of the twentieth century was that scientists could reproduce the petroglyph images in the plasma laboratory. The laboratory plasma that produced petroglyph images was the type of plasma that creates the Earth's auroras. Under certain conditions, those images drawn in petroglyphs could have appeared in the auroras during ancient times. Numerous areas through

FIGURES 1.13 and 1.14. The narrow Devil's Den cave entrance is blocked with rock debris. There is an unusual cut in the stone, leaving the entrance that seemingly drains water down the hillside. Theories of what happened here include that it was once sand on a seashore that turned to rock, and that the stone was melted by a plasma outburst or coronal mass ejections (CME) from the sun. Other theories are welcome to solve this mystery. World renowned geologist Dr. Robert Schoch wonders if some rock art petroglyphs are eye-witness accounts of global plasma events in the ancient past, based on Dr. Jeff Ransom's work.

FIGURE 1.15. The strange cut down the center of this image is intriguing. Note the huge arrowhead on the left of the cave entrance, which was observed by former owner W. L. Spoon in his "Gem of Scenic Beauty" pamphlet in 1929. Is that rock art or another part of the natural versus man-made theme? It looks a lot like Harrison's Signal Rock, mentioned earlier in this chapter.

> the world could have seen similar images in the sky during the same time frames.
>
> —Dr. Jeff Ransom (2018)[30]

There are currently three possible caves that fit the description of the Devil's Den cave. We spent some time at each cave, but all were filled with rock debris and didn't allow further exploration. The cave debris raises questions about whether it's natural versus man-made yet again. None of our four intrepid young people felt air movement in any of the cave entrances on the day of our visit. It was at the last spot on the south face that most closely matched the newspaper descriptions. Interestingly and mysteriously, the master number eleven (an

30. Jeff Ransom, *Auroras, Petroglyphs and Pagans* (South Carolina: CreateSpace Publishing, 2018).

important number in numerology), or two columns (important symbol in Masonic Freemasonry), was clearly carved deeply into the stone at the entrance. It felt right. There was a small entrance to shimmy into, but the larger area beyond that was blocked with stone debris just like the other entrances. Then Kenny found a feather right in front of that cave opening, which gave us further confirmation that we'd found the spot.[31] Finding a feather right on your path is a confirmation you are on the right path in the oral histories of the planet.

A few months later, on a return visit to the mountain, air was felt at this cave entrance as rain approached. The movement of air just might have to do with atmospheric conditions, just as Michael Smith said. The day of our hike with the Beasley grandchildren was cloudless and under a high-pressure system.

This trail around the Devil's Den is the trickiest part of the hike. It has the texture of molten sand, which doesn't allow for good footing. It is also slanted downward with a sudden drop-off down the steep side of the mountain. Before we moved back toward the saddle area, we all paused to look at the curious straight line cut into this sandy-looking rock coming out of the Devil's Den cave. That straight line cut in the stone discharges into the forest below. It is like a small, dried stream of water flow with enough power to cut a trench, but the trench sides are too square to be natural. Perhaps former owners Spoon or Beasley made it to keep it from being too slippery there? Later, I found an image of the Devil's Den in an old edition of the *Winston-Salem Journal*. It confirmed our suspicions. We were right about this being the Devil's Den area and cave.

At this point in the hike, we came to a possible significant archeoastronomy spot on Pilot Mountain, although there is a mountain of research to be done to confirm it. There appears to be a solstice and equinox sun dagger, much like the one at Chaco Canyon in New Mexico, with two different access points for the sun's rays. Another area that bookends this spot is on the other end of the saddle area. It also has potential for some archeoastronomy discoveries on the same east-west axis on the ridge (see Fig. 1.16). Both are high above the trail that bookends the saddle area. A third item to note is a rather interesting piled stone altar area that, while not looking natural, would have been quite an engineering feat to construct on the side of the mountain.

One of the three largest stone faces on Pilot Mountain State Park found so far is in this area but, going clockwise around the Big Pinnacle, you have to turn around to see it from a specific spot looking east.

Remember, discovery is more than half the fun on Pilot Mountain.

Of all the stone faces on the Pilot Mountain trails, people tend to look at this one near this altar area the longest. Some have an expression indicating, "that can't be." Almost all the images of the old stairway to the top of the Big Pinnacle feature a stone face to the left. It is subtle but looking closely with soft eyes, you'll see it. As we looked forward, moving toward the Little Pinnacle from the Big Pinnacle, we saw what appeared to be a whale head and a full-size dolphin near

31. There are many Native American legends and lore on finding a particular feather on your path for getting your attention.

FIGURE 1.16. This is one of many unusual spots seen at Pilot Mountain State Park in the natural versus man-made theme. Is it an old fort or cairn, or just natural? There appear to be many well-placed spacers, or rock shims, supporting the slabs of rock that are better seen in person.

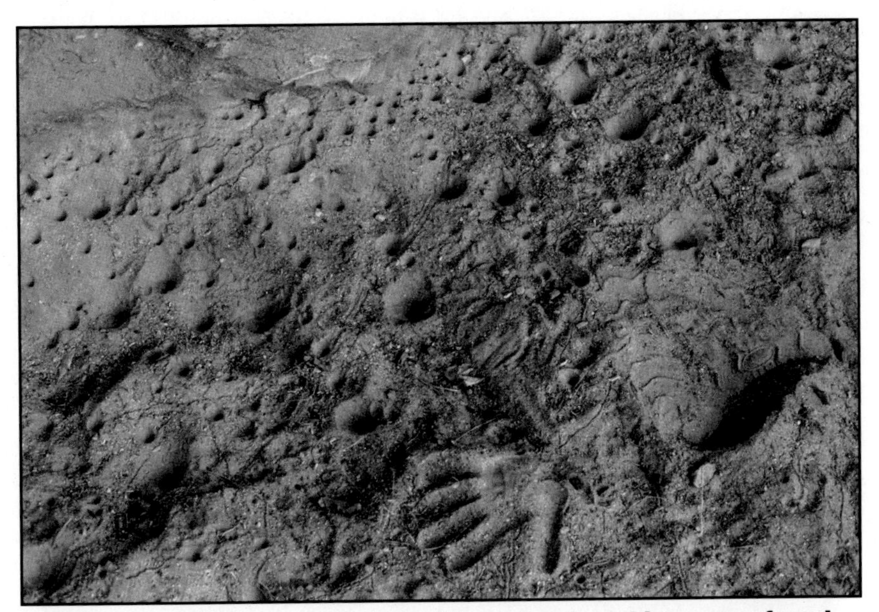

FIGURE 1.17. These unusual, three-dimensional markings were found in the sand on the Pilot Knob Trail one afternoon. These appeared to be poked out rather than into the sand in the form of circle and trails between them. The rock art style handprint in the sand poking out was also very unusual.

where the whale's blowhole would be. Sort of like the dolphin was giving the whale a little nudge. These stone animals are visible all year round but are best seen when the leaves have fallen. If you find the whale and the dolphin, you are very close to where the stairs and ladders used to be to climb to the top of the Big Pinnacle.

CLIMBING THE BIG PINNACLE

> The climb is made in two stages by the hardy. Automobiles halt on the shoulder, and a foot trail lead through the rough ravine, amid tangled rhododendron, laurel and chestnut oaks, to the base of the towering knob.
>
> A sturdy staircase (106 steps) leads upwards—though at a steep angle—and the view from the top makes the climb a modest investment. Mr. Beasley's somewhat fulsome literature speaks of 'Nature's freak,' and 'primeval forest,' and 'veritable flower garden,' but a visitor on the knob is apt to forgive him this, with the farms of three counties spread at his feet, and the vistas unbroken by other peaks, as is usual in mountain country.
>
> It is one of the state's most accessible mountains, so far as the bulk of the Tar Heel population is concerned, but despite its unique charms, it remains one of the least familiar of the state's scenic attractions, perhaps for the very reason that it seems hopelessly misplaced in the open country of Surry, and that the view from its hard-won summit is, until the last upward step, incredible.
>
> —Burke Davis (1940s)[32]

"I was up there maybe two or three times after I was out of school," Beverly fondly remembered when you could go to the top of the Big Pinnacle on the wooden stairs. "The beautiful trees up there. It's kind of sparse. You know, it's not really. . . it's vegetated, but it's not heavily vegetated. It has got a US Geological Survey brass marker on top of it from sometime in the '50s."

The original stairs were built in the same location where people used to climb up the rock face of the Big Pinnacle. This was also the same area where the ladders were placed. The stairs were built in 1929 out of chestnut wood with lock anchors and balustrades (a row of small columns topped by a rail). That assured rigidity and safety for the climbers on the stairs. Just two years after the mountain

32. Burke Davis, "Pilot Mountain's Knobby Summit Has Been a Landmark for Centuries," *Greensboro Daily News*, December 26, n.d.

became the Pilot Mountain State Park, state inspectors condemned the forty-one-year-old steps, and they were destroyed.

After finding the Devil's Den with the Beasley grandchildren, our next stop at the campground area brought back a flood of memories from their youth of working for their grandmother on Pilot Mountain during the summer. J. W. and Pearle Beasley had five daughters and a son. Their one son died at the age of five. There were thirteen grandchildren. Their third daughter, Ruth Carole Sperry, would help run the mountain during the summer when she wasn't teaching at the Pinnacle Elementary School and later the Pilot Mountain Middle School after J. W. Beasley passed. Some of the other Beasley "sisters" would fill in for her when needed.

Before we head back down the mountain, let's see what Carole B. Sperry wrote about Pilot Mountain State Park:

> In May it is one gorgeous bed of rhododendron and mountain laurel. It is a carnival of wild flowers. Those fortunate enough to see it then will never forget its beauty. Even the parapeted pinnacle is studded in crevices with beautiful bouquets far out of reach of man's hands. The flowers are most abundant on the north or shaded side of the Pinnacle. On the mountain you will see the shrubby bear oak (Quercus ilicifolia) – one of the five species of pine is pinus pungens, the table mountain pine, a small tree of irregular shape with curious prickly cones.
>
> Mosses and lichens grow on the rocks, small polypodies and spleenworts in the crevices, and where enough soil has collected there are carpets of

FIGURE 1.18. An example of Mountain Laurel found on the trails of Pilot Mountain State Park in May.

> galax and trailing arbutus. John Wesley Clay once wrote that about this time, 'each spring I have the urge to go to the mountains and get the first glimpse of the beautiful trailing arbutus.'
>
> The sweet fern can be found on the mountain. It really is not a fern, but a shrub, Comptonia peregrina, belonging to the barberry family. Its dark aromatic leaves are deeply cut in a coarse sawtooth pattern. One can find kalmia also peiris floribunda, a beautiful shrub. It is found a very few places in NC.
>
> —Carole B. Sperry (1983)[33]

PILOT MOUNTAIN POOL AND DANCE AREA IS NEXT ON THE GUIDED TOUR

We returned to our cars on top of the mountain. This could have been the end of our guided tour, since two members of our party got in their car and left for the airport. However, the rest of us still had plenty of energy; our guides knew exactly where they wanted to go next and there was no hesitation. After dropping off our two companions headed for the airport at the visitor center, we headed back for the campground at the base of the mountain. We were now a party of six. We all set our sights on finding the pool area, but where was it hiding all these years later? This was really getting fun, because I had read dozens of accounts about the pool but had never being able to find the location. Old roads that have been out of use for fifty-five years have crumbled and are now walking paths. Having gone back and forth in this area near the old pool location, we were happy to have guides now. We noticed the new wooden benches built after the state park system bought the mountain from the Beasley family. These benches might be for Sunday worship or for hearing a talk at the campground.

Since nearly everything is gone above ground from this formerly popular area, the grandchildren had to reorient, turning around looking for something, anything, familiar. The dances, they recalled, drew three to six hundred people every Thursday night to hear the local band, the legendary King Bees, led by multi-instrumentalist Billy Long. They had a red-haired vocalist that some might remember, named Dale Riddle. Everyone who speaks of the King Bees gets a big grin on their faces, including the Beasley grandchildren.

Once in the general area of the dances, near the current campground wooden benches, some of Pearle Beasley's stone walls were found. That was sort of the Rosetta Stone for the Beasley grandchildren and anyone else who used to attend activities there to orient themselves. Soon after, the earthen dam for the mineral

33. Carole B. Sperry, "The Story, Legend and Facts about Pilot Mountain," in *The Heritage of Surry County North Carolina, vol. 1*, ed. Hester Bartlett Jackson (Surry County Genealogical Society, North Carolina: 1983), 629.

FIGURES 1.19 and 1.20. Old bathhouse and pool with a beautiful view of the Sauratown Mountains in the background. Both images courtesy of the Frank Jones Collection NC Room at the Forsyth Public Library.

spring, which held back millions of gallons of water at one time, was found mostly intact. Just down the hill from the earthen dam is where the pool is now buried. It will be a huge archeological discovery in a few thousand years. Between the two locations, the old stone BBQ grill built in the 1940s was also found in a beautiful lush green grove of trees.[34]

"Grandma was a supervisor and a pointer," grandson Kenny recalled with a grin about all her stone walls standing in the area near where the old snack bar used to be. "On top of this stone wall you can see people could sit with their towels and wait for their parents to pick them up."

34. Staff writer, "Improvements Begun at Pilot Mountain," *News & Observer*, April 11, 1946.

FIGURE 1.21. Beverly Glace and Kenny Glace Jr. in the pool and bathhouse area at Pilot Mountain on our hike in 2020.

There are more stone walls up near the foundation of Bert Coleman's former home, near the stone building off the main road. Beautiful yellow daffodils continue to bloom year in and year out in the Bert Coleman garden. The historic Coleman home was demolished decades ago, just like the stone Gate House. Another home on the mountain, where Bert's son and daughter-in-law, Ben and Winifred Coleman, lived, has also been destroyed. Why those two historic homes were destroyed is a mystery. Just up the path from the Coleman foundation, there is another large mineral spring. The historic Gillam Hotel, which was run by a former owner, was built in 1830 and still stands on the other side of the mountain.

Beverly pointed to the heavy rocks on the old earthen dam placed there three-quarters of a century ago. We hiked up the side of the dam to get a better look. The mineral spring that used to feed the holding pond, which led to the pool, now flows to one side. There is barely any water, unlike when it was built in 1946. There used to be millions of gallons. It still looks to be deeper than the fifteen feet behind the old dam. Currently, the rushing water from storms revealed a tire, some rebar, and metal poles sticking out of the dam. At first, they thought that a natural filtering system of sand and gravel would be enough to clean the pool water, but the overwhelming success of the pool made that unworkable.

"There was a chain-link fence all around the dam here, and then the pond below the pool also had a chain-link fence for safety," Beverly, perched at the highest point of the old dam, pointed at each place in turn. The old pool below the pond had a slide for kids plus a low and high dive. Some people came to just watch the high-dive action at the pool sitting on a little hill above the pool.

The dam was built up above to provide water to the pool and came right down. They initially set up a filtration system that was sand and gravel based. It didn't work.

So, I'm out there at five o'clock in the morning with a pump and vacuum out in the pool, vacuuming all that stuff off the bottom. It was cold, so cold. You had to get right in and do it. Yeah, but one day, I made a mistake. I put my nose over that bucket and those chlorine vapors that'll burn your lungs—*yeah, no kidding.*

—Kenny Glace (2020)[35]

Next, they told us where the bathhouse and refreshment stand were in relation to the rock wall we were at nearby. Memories of the pool days all came rushing back. They told us stories about how the storms came right over the mountain from the southwest and often you wouldn't see the dark storm clouds coming. Beverly used to work at the old snack bar and we could imagine it as she described it. We could almost smell the hamburgers cooking. It was clear that this place was full of fun and frolic.

"We sold snacks and souvenirs," Beverly explained. "In the summertime, we started in the morning; we would go just buy tons of ice and we'd get these coolers. A couple of us would go and take the drinks and the snacks. We'd sit at the top and sell drinks. It was kind of like a drive-in movie kind of place where you could get a hot dog or a hamburger. The best chili hamburgers that you've ever put in your mouth. We had a little room and had everything to fix everything. It was just primitive, right? We had a little freezer, and it was a lot of fun. It wasn't like any professional kitchen at all. Back in those days not much water was sold."

"You didn't buy water back then," they both chimed in at the same time shaking their heads no.

R. M. Collins was the head lifeguard for the pool, and he came up with the idea for the dances with the King Bees' music. That turned out to be a brilliant idea for Thursday nights and a lot of fun, but added to Kenny's duties. Every Friday morning, Kenny had to clean the parking lot of cans and bottles before the first swimmers arrived.

"I was a lifeguard for two summers with R. M. Collins," Kenny remembered working summers for his grandmother, puffing out his chest while still looking rather sheepish. "I was up there and we're having a thunderstorm going on. Of course, everybody really cleared out to the bathhouse and over to the pavilion. So, I'm sitting there, smart me, washing dishes and at the same time a lightning bolt hit nearby. The next thing I know is I was lit up—oh man! I was touching the metal probably."[36]

"We had baskets and you get a pin, and we put all your stuff in the basket. Then you bring it back to the desk and that's where you store all your stuff like a coat check," Beverly explained about the bathhouse. "It was a stainless-steel pin and that was the locker system."

Swimming lessons in the icy-cold mineral spring water were a pretty good deal at the Pilot Mountain pool for years. For just $2.50, you got lessons Monday through Friday for six weeks. You had to get up early to catch the bus from town,

35. Glace, Glace, and Sperry, interview.
36. Glace, Glace, and Sperry, interview.

FIGURE 1.22. Bert Coleman's home foundation now near the Drink House. Nearby, the trail to the Pilot Mountain Visitor Center goes right by it near a large mineral spring.

which usually arrived at the pool at 8:30 a.m., and the lessons would last for a couple of hours. This was the time that cars would drive up and down the entrance of Pilot Mountain to see what was going on and to be seen. Outside the pool fence, people would sit around socializing, watching the high divers in the pool or watching the rather regular rescue by the lifeguard of someone who didn't know how to swim. On Sunday afternoon, the area in and around the fenced-in pool was packed.

"If you wanted to go swimming, you could never just walk into it, you just had to take the plunge because it was cold," Beverly with a huge grin. "You knew when to come out when your lips turned blue on a hot summer afternoon."[37]

When Pilot Mountain was purchased in 1968 to become part of the North Carolina Park Service, the twenty-year-old pool was broken up and filled in with dirt because it was too close to the road. That road was going to the new campground. Plus, the pool might have been in need of repair after two decades of heavy use.

"They've completely covered in the pool. They filled in all the ponds, they had the dam up above and that's gone," Beverly remembered. "The pond below the apple orchards is gone. Now if you really want to find the apple orchard, just go where morels grow because they love dead apple trees. If you ever go up on that piece of the property you will find just a crapload of morels. Then you will know that's the old apple orchard. Every year we would go harvest, oh, five, six, or seven bushels of apples. They were the best apples in the world. I know we would get these big bushel baskets, I was really young at the time, but I would love to go

---

37. Glace, Glace, and Sperry, interview.

FIGURES 1.23 and 1.24. These images were taken from the same location near the Pilot Mountain High School decades apart. Another example of how quickly a landscape changes in a lifetime. Image on top courtesy of Frank Jones Collection at the Forsyth Library NC History Room in Winston-Salem.

because we'd get to climb trees. They weren't the prettiest applies to look at but, let me tell you, they're the tastiest apples you ever ate."[38]

38. Glace, Glace, and Sperry, interview.

"I remember picking bushels of apples there and then taking them to my grandfather Spoon's home to place them in his basement," according to Frances Alexander Campbell, W. L. Spoon's granddaughter, in a phone interview earlier in the springtime of 2020. "That apple orchard was near the pool area but now is gone."[39]

In a few decades, the stories from this chapter will be forgotten and without the eyewitnesses to prove them. You will have to dig, literally, to prove many of these stories in a just a few years. The doubt about the airfield on top of Pilot Mountain is a perfect example of the doubt some might harbor regarding these stories' existence. We will let our ancestors speak on that in later chapters.

> There's a story for almost every mile of the Blue Ridge Parkway. Somewhere up on the Blue Ridge is where the Great Wagon Road came down the mountain in sight of the Pilot Mountain landmark.
>
> The road itself is new but the land it spans is old and so are the tales that whisper from its cloud-hidden coves and laurel-crowned mountains. The ghosts of old Dan'l Boone and 'Chucky Jack Sevier' hover in its mile high passes. The dead ashes of Indian campfires lie under its black surface. Old-timers, horse-and-buggy pioneers, know the lore of the road and if you'll stop and tarry a bit with them, they'll tell you what took place on yonder mountain, or in the pass up ahead, or in the valley just there beyond the river.
>
> —John Parris (1959)[40]

If only all archaeologists or Scots-Irish history buffs, like myself, had two guides as informative as the Beasley grandchildren were this last day of summer. Then our past human history would be neither forgotten nor repressed.

"If you want to get a beautiful view of the Pilot Mountain or the whole Saura range, I have a location for you to go," Kenny suggested for a spot to end the day after our late lunch. "You go through Pilot Mountain on old 52 and go to Cook School Road. Okay, and take a right. It's back there about a mile and a half and make a left, and it's just a ridge up there. You go to the top of that ridge."

After that lunch, just before four in the afternoon, we ended our day together about eight miles from the mountain on an overlook to the northeast that both our guides thought was special. It was a beautiful view of the entire Sauratown Mountain range with Pilot Mountain to the west like a stack of chimneys, just as William Byrd mentioned in 1728! We parted for our homes with full hearts.

---

39. Frances Alexander Campbell (granddaughter of W. L. Spoon), phone interview with author, July 3, 2020.
40. John Parris, "Roaming the Mountains: High Road of Adventure," *Asheville Citizen Times*, June 26, 1959.

FIGURES 1.25 and 1.26. Later, even more guided tours with Beasley family members followed over the years. One time, Patricia Sperry brought along a principal (above) of one of the schools in Boone, NC. On another one, we even got a tour of the inside of the old Drink House (below) with one of the park rangers, who was taught by Carole B. Sperry at the elementary school in Pinnacle, NC . . . small world.

FIGURES 1.27 and 1.28. Inside and outside the Drink House during Beasley ownership of Pilot Mountain. Both images courtesy of Patricia Sperry.

We will now examine the historic times of Pilot Mountain starting in 1753 and continue on to it becoming Pilot Mountain State Park. In the next book in this series, we will concentrate on what is known of the prehistoric times of Pilot Mountain and the early maps made of the area around Pilot Mountain.

PYRAMID ASPECT

Two university professors in different centuries took note of the mysterious truncated pyramid aspect of the mountain, which is now only seen from country roads away from the mountain and not the US Route 52 . . . hidden in plain sight, it would seem. The truncated aspect of the mountain with the flats is a loud bell ringer for mound experts.

> In the first glimpse we catch of the Pilot in Rockingham, it resembles a magnificent temple with a superb cupola, not unlike the picture of St. Peter's at Rome. The uncommon symmetry of its structure is preserved on a much nearer view. Nothing could exceed the regularity and beauty of its appearance, as it presented itself to President Caldwell, Professor Andrews, and myself, on a summer evening of 1823, as we were approaching it from the east, a little before sun-set. Its dark side being towards us, we could more distinctly observe its finished outline, which was still illuminated. The figure now presented by its sloping sides and perpendicular summit, was that of a triangle, having a portion of its vertex removed and replaced by a parallelogram; while trees and shrubbery that graced the outline, appeared like delicate fringe projected on the western sky.[1]
>
> —Professor Denison Olmstead
> University of North Carolina at Chapel Hill (1823)

> The Area: Pilot Mountain is the most southwestern outpost of the Sauratown Mountains and is located in the southeastern corner of Surry County, North Carolina. Several miles separate it from the two nearest spurs of the Sauratown Range. Except for hills in the immediate vicinity, the 2413 foot peak (Pratt 1917) is isolated and is surrounded by the characteristic gently rolling topography of the Piedmont Plateau.
>
> Topography: The mountain may be separated topographically and vegetationally into two distinct parts. The eastern half is an almost perfect pyramid, which is topped by the Knob, or Big Pinnacle (fig. 1); the western half is an elongated gently rounded ridge, which, extending westward from its highest point, the Little Pinnacle, is outlined by the Ledge. Both the Ledge and the sides of the two pinnacles are nearly vertical cliffs composed of horizontally stratified rock.
>
> On the north the two sections of the mountains are separated by Grindstone Ridge, which slopes gradually downward from the southeast corner of the Knob, and several shorter ones are found on all exposures. Northwest of the Big Pinnacle, separating it from the Little Pinnacle and from the Grindstone Ridge, is a broad cove, which is dissected by several shallow ravines and broad, low ridges. A number of ravines occur on the eastern half of the mountain, but few are found on the western half.
>
> There are several flats at various altitudes on the eastern pyramid. The two largest, both at 1600 feet, are Hickory Flat, which extends across most of the east slope, and Poplar Flat, which covers fifteen or more acres on the south slope.[2]
>
> —Ruby M. Williams and H. J. Oosting
> Duke University, Durham, North Carolina (1944)

1. Denison Olmstead, "The Pilot Mountain," *Raleigh Register* and *NC Gazette*, November 3, 1826.
2. Ruby M. Williams and H. J. Oosting, "The Vegetation of Pilot Mountain, North Carolina: A Community Analysis," *Bulletin of the Torrey Botanical Club* 71, no. 1 (January 1944): 23–45, https://www.jstor.org/stable/2481485.

FIGURE 2.1. Stone face on a Pilot Mountain State Park trail.

2

Arrival of the Moravians Near Pilot Mountain in 1752

FIGURE 2.2. This is a look from the south side of Pilot Mountain State Park with an old tobacco barn on the right. Courtesy of the Surry County Historical Society.

> Although the Saura people abandoned their villages, the mark ancient Native American tribes left on the region remained. The paths used by ancient Native American tribes became the Great Wagon Road, used by 18th century colonist to travel and settle further South. On the Great Wagon Road, Jomeokee or Pilot served as a landmark for German Moravian settlers traveling to their new home of der Wachua, or Wachovia, present day Winston-Salem. Wachovia quickly became an established urban center and led to further settlement of the Piedmont region including the Town of Pilot Mountain.
>
> —Town of Pilot Mountain (2022)[1]

1. "History," *Town of Pilot Mountain*, November 8, 2022, https://www.pilotmountainnc.org/community/page/history.

The Moravian journals that focused on North Carolina in detail started in the 1700s. These important historical records were written between twelve and twenty miles from Pilot Mountain State Park as the raven flies. The aim was to write on a daily basis, and this was accomplished. These accounts are important descriptions of the Pilot Mountain neighborhood for nearly the past three centuries. There are Cherokee Myths from the Oral Tradition documented by James Mooney and Thomas E. Mails that allude to the Piedmont area, but the Moravian journals are a historical treasure within sight of Pilot Mountain. The Moravian journals contain written entries, later translated from their original German, by one to three individuals each day. Their aim was to write detailed descriptions of daily life to share current events with others spread across the planet. The journals were also documented so future generations could understand their history. They include accounts of early life as explorers in the back country of North Carolina. There are many references to visiting Pilot Mountain.

The Moravian Church is over six hundred years old. It was formed when members broke away from the Catholic Church because they wanted Sunday worship spoken and the Bible written in their language. This split was one hundred years before Martin Luther nailed his letter on the church door. The Moravian Church is a denomination within the Protestant religion and they do share the same core beliefs. They have a strong pacifist stance that was tested to the limit during time of war right on their land for multiple generations. The Moravians paid a heavy price for being a pacifist in a time of war time and time again. Was this historical record of their pacifist stance something that helped strengthen their fortitude against killing another human being? Did the oral history of the Cherokee and the written journals of the Moravians create a soul connection between the two? Did one "principled people" recognize that in another from their first meeting within twenty miles of Pilot Mountain?

Edward Rondthaler remarks on the importance of the Moravian journals in terms of the history of the State of North Carolina in the first book to be translated from German to English:

> Owing to their worldwide connections in the Unitas Fratrum, and the scholarly methods of their leaders, they brought with them the habit of keeping precise records of all current events. Copies of these [journals] were communicated to their Brethren in other parts of the world, and the originals were carefully deposited in their Archives, now at Winston-Salem. In the later colonial years of North Carolina's history, the contemporary accounts were sparse and unconnected; there were many breaks and gaps in the story of the State. The Moravian Records are perhaps the only consecutive historical account which North Carolina possesses for the critical years of her development. The Moravians were acute and watchful [analysts]. They recorded not only the doings of their own religious body, but made note of the state of the weather, incidents

of travel, prevailing fashions, features of topography. They mentioned the many distinguished men of the State who visited them, and whose descendants are a valuable element in our population to this day. These accounts are given in the Moravian Diaries, in Travel Diaries, and in the 'Memorabilia,' as they are called, which are the accounts of the successive years from 1753 to the present time.

—Edward Rondthaler, D.D., LL.D. (1920)[2]

THE JOURNALS OF THE MORAVIANS

When visitors came to meet with the Moravians, often for their very first excursion in the area, they wanted to hike up Pilot Mountain, which they had read about in the journals. I have attempted to piece together everything said about Pilot Mountain based on the index of thirteen volumes. Since these volumes cover nearly six thousand pages in total, some might have been missed but the flavor of those times is captured in this chapter.

At first, only a small group wrote daily reports on the selection of the one hundred thousand acres they bought from Lord Granville in England. Land ownership in North Carolina in the 1500s, 1600s, and 1700s is convoluted and would require an entire book to sort out. These daily journal accounts describe travel coming down the original Great Warrior Path from the Philadelphia area, scientific area weather reports, pandemics, vaccines, garden building in a forest, creating a grist mill, grind stones from Pilot Mountain, the French and Indian War, the Revolutionary War, the War of 1812, and the Civil War. Sometimes their own backyard fencing was used to build fires to keep the troops warm on streets in front of their homes. Repeatedly, the Moravians found themselves "in the very theater of war" as pacifists. One side or the other would label them as spies in times of war.

Strong earthquake reports from 1811–1812 (New Madrid, Missouri) and 1896 (Charleston, South Carolina) record many aftershocks felt in the area around Pilot Mountain for months. Area newspaper accounts also documented these large North Carolina earthquakes both within the state and from elsewhere.

The history of the Moravian Church has been so well documented over the centuries in journals they wrote all over the world. The first journals started when they left Bohemia to go to Germany, where they found a benefactor to cross Europe, then England, the Caribbean, and on to the colonies, where they settled in Pennsylvania among other places in the world. Their journals were originally written in German, later translated to English, and published from 1922 to 2006 so far. The generous and much-appreciated permission for the use of these

2. Adelaide L. Fries, Records of the Moravians in North Carolina, vol. 1, 1752–1771 (Raleigh: Edwards & Broughton Printing Company, State Printers, 1922), 5.

journal entries for this chapter were granted by Moravian Archives, Southern Province in Winston-Salem, North Carolina.

The Moravians were industrious and, early on, cleared good North Carolina trading paths. Later, there were roads leading right to them to help sell their goods and a wide variety of services. By the 1800s, these roads would be known as the Main Great Wagon Road, Hollow Road, Virginia Road, Richmond Road, Dan River Road, and Town Fork Road, all within sight of Pilot Mountain, which was used as a landmark if you got to the top of a hill or clearing in the deep virgin forest.[3] Early visits to Pilot Mountain were on the Indian trails and on those Moravian cleared roads. Ironically for these pacifists, these early trails and roads also provided easy access for the military troops during the wars, leading right to them.

Like so many who came to the backcountry of North Carolina, they did not arrive from the coast. The early settlers came down the Great Warrior Path. Today, as you pass a thick forest, you can't see much more than ten feet in. You may pause to think about turning your car, with everything you own, into that forest to travel forward along a deer trail three feet wide. What would you have to do to make headway with your car in that thick forest? After all that work in the forest, you come out of it and gaze upon a creek, stream, a river, or a mountain. What do you do then? Instead of a car, you had a team of horses and a wagon with banks on both sides of the river or creek to deal with.

MORAVIAN'S EXPLORE NORTH CAROLINA BEFORE THEY SETTLE

In the 1700s, Moravians first traveled from North Wilkesboro to Hickory to Salisbury, but settled in Bethabara, which is now a historic district in the outskirts of present-day Winston Salem. Bethabara is separate from Old Salem, but both are well worth a visit.

> Receiving the grant of land from Lord Granville did not remove all the difficulties in connection with the founding of the proposed Moravian settlement. Little was known of North Carolina by the average inhabitant of Europe. The very terms of the boundary description impress us with the vague idea they had of their possessions in America. After seven of the eight Proprietors who owned the American soil from the Virginia line to a point in Florida had relinquished to the crown their rights, Edward, Earl of Clarendon (Lord Granville), retained his portion. This territory of Lord Granville extended from the Virginia line to a point about seventy miles south, and according to the terms of the deed, from the Atlantic Ocean on the east to the Pacific Ocean on the west, or, as the

3. Fred Hughes, *Surry County Historical Documentation Map* (Jamestown, NC: The Custom House, 1988). Surry County Historical Society.

FIGURE 2.3. A recreation of the Wainwright Barn at Bethabara Historic Park, which is a separate location from Old Salem.

> Spangenberg papers describe it, 'to the South sea.' Thus, when the Moravian explorers began their journey, they had a strip of land seventy miles wide and three thousand miles long from which to select their tract.
>
> —John Henry Clewell, Ph.D [4]

The following is an excerpt from the Moravian journals during their explorations down the buffalo trail of the Great Warrior Path. To some, this Piedmont backcountry of North Carolina was thought of as the fourteenth colony because it was so separate and different from the rest of the colonies.

> **October 30th, 1753.** The weather was bad, it rained and snowed, but we kept fairly dry under our tent. Our horses had strayed off and it took several of the Brethren nearly all day to find them, and we were glad when we had them back for, we had heard that in this neighborhood horses were often stolen, and that might have happened to ours. As the Brethren came in cold and wet through and through we had a cup of tea all round, and enjoyed it together.

4. John Henry Clewell, *History of Wachovia in North Carolina* (New York: Doubleday, Page & Company, 1902), 4.

October 31st, 1753. Next day the journey was continued over the frozen, snow-covered ground. 'The farther we went the more snow we found, and travel was difficult,' for the 'Upper Road' which they were following lay along the hills of the Blue Ridge, and sometimes 'the road sloped so that we could hardly keep the wagon from slipping over the edge of the mountain, and we had to use the tackle frequently.' Often too they had to cut down trees to let the wagon pass, or work the road before they could cross a particularly bad place, sometimes even cut out a new track to avoid an impossible mud hole. But courage, perseverance, and hard work won through, and at last on Nov.7th from the top of a little hill they 'saw the Pilot Mountain in North Carolina, and rejoiced to think that we would soon see the boundary of Carolina, and set foot in our own dear land.'

November 11th, 1753. On the 11th they met a man from North Carolina, who lived not far from the Wachau, and 'he told us that it is generally known that we will soon arrive, that he had heard that we had two ministers with us, which was a good thing, for the people lived like wild men, never hearing of God or His Word.' He was also glad they had a doctor with them. All along the way people had welcomed the advent of a minister, and quite a number of requests were made for the baptism of children, - which the Brethren did not feel themselves at liberty to grant, - while others asked that on later journeys visits might be paid them, and services held.[5]

The experience of the first Moravians who "rejoiced" at seeing the Pilot Mountain landmark in the distance would be repeated by wagons coming down the old Great Warrior Path time and time again for decades. Later, wagon trains coming down the same way but calling it the Great Philadelphia Wagon Road had the same experience. The wagon master might have mentioned it several days in advance to encourage the weary travelers. Generally, the first people exploring this area to settle came after the fall harvest, which financed their trip to a new land with everything they owned and needed upon arrival. When they found desirable land and shelter, the spring planting was foremost on their minds. When looking for new land, you wanted to avoid areas with river cane. This was very abundant in what is now the southeastern United States and is like a smaller version of bamboo that grows like a weed. River cane can grow twelve feet high and is difficult to clear. The Cherokee weaved all sorts of baskets and containers with the cane. There is a set of cane baskets from North Carolina in the British Museum that were brought to London in 1725.

5. Fries, *Records of the Moravians*, 77–78.

> There are no canes, nor any sign of them, but plenty of grassland. Corn, wheat, oats, barley, hemp, etc., will grow here, of wood there is no lack; we have included in our tract a beautiful chestnut forest, and fine white pine. The water is clear and delicious. Among the various stones there is a variety which Br. Antes thinks the best for mill-stones that he has seen in America. Many hundred, or thousand, wild apple trees, 'crab trees,' grow here; probably vinegar and spirits could be made from them.[6]

Had these wild apple trees been tended orchards by the Native American tribes who lived in this landscape for centuries? The Saura tribe had been gone for over fifty years by the time the Moravians moved to the area. There was even an apple orchard at the base of Pilot Mountain, near the modern-day camping area, when it was privately owned. Members of the Spoon and Beasley families commented on how ugly the apples were from Pilot Mountain but how great they tasted.

Sometimes fording a river required taking the wheels off a wagon and using block and tackle to get from one riverbank to the other. A steep riverbank on either side sometimes required hours of digging and engineering on the spot. Many who didn't know how to swim would drown crossing large rivers and be buried there. As travelers went down this woodland path in those early days, they carried the current news of the day. At every stop, people would be eager for news of the outside world or from home in Ireland, Scotland, France, or Germany, as they might not have heard anything for months or even years.

THE FIRST MORAVIAN SETTLERS

> **1753.** The following are the names of the nine Brethren, who arrived as first settlers:
>
> John Beroth, farmer, from the Susquehanna, Pa.
> John Lisher, farmer.
> Herman Lœsh, miller, from Pennsylvania.
> Jacob Lung, gardener, from Wurtemberg.
> Christopher Merkle, baker.
> Erich Ingebresten, carpenter, from Norway.
> Henry Feldhausen, carpenter and hunter.
> Hans Peterson, tailor from Denmark.
> Jacob Pfeil, shoemaker, from Wurtemberg.[7]

6. Fries, *Records of the Moravians*, 56.
7. Levin T. Reichel, *The Moravians in North Carolina: An Authentic History* (Philadelphia: J. B. Lippincott and Co., 1857), 200, appendix.

FIGURE 2.4. View out the old-style window that might make you wonder, friend or foe or wild animal?

The Moravians stayed in an abandoned shack with a leaky roof at first, but then they constructed sturdier buildings and grist mills. They also cut down the huge forest around their dwellings to garden and plant food crops. Early on, Cherokee neighbors met them and they got along well from the very start. The Cherokee were invited to eat and they accepted from their very first meeting.

In one journal entry, a Mr. Shepperd is mentioned, who claimed the top ten acres of Pilot Mountain, which he was granted years later on July 31, 1778. Shepperd previously owned this land on Pilot Mountain when England owned it, paying taxes to the crown.

SUMMARY OF THE FIRST FEW DECADES OF THE MORAVIAN SETTLEMENT

Here are more Moravian journal entries that provide a summary of this time. Later, we will get into the detailed journal entries, which are just as fascinating.

1757. Among those coming to the Bethabara mill are mentioned Mr. Shepperd and Mr. Banner.

1760. Two hives of bees were brought from Tar River, 120 miles, which increased very fast; in consequence, many bears made their appearance in the Fall. In December, immense quantities of wild pigeons made their appearance and roosted nearby for nearly a month. When together, at night, they covered only a small tract of woods, but were clustered so thick upon the trees as to break down the largest limbs by their weight. The noise made by them in coming to their camp at night, as well as the fluttering, &c., during the night, and their breaking up in the morning, was heard at a considerable distance. The spot was marked for many years.

1761. January very cold, and thick ice on the mill pond, strong enough to drag heavy logs over it to the saw-mill.

1763. In Bethabara and Bethania wells were dug, and the first pumps introduced into this part of the country.

1765. John Leinbach, with his family of seven children, arrived from Oley, Penn., and bought lot No. 1, the so-called Leinbach tract.

1767. The County Court in Salisbury gave permits for three public roads, one leading from Salem to the Town Fork and Dan River, another to Belew's Creek and the Cape Fear Road, and the third southward to the Uwharree.

1769. Great abundance of wild grapes; nineteen hogsheads of wine were made in the three settlements.

1770. Abundance of caterpillars, which destroyed much of the grass and grain. The place for the burial-ground of Salem was cleared and fenced in. Roads opened to Salisbury and Cross Creek.

1771. Much harm done on the corn by the squirrels, also many bears in the woods.[8]

To be clear, the Germans weren't the only group to settle in the area around Pilot Mountain.

German settlement of the region continued, along with settlement of the Scots-Irish, and English Quakers, who relocated due to the scarcity and expense of fertile land in the northeast. With these settlers came a lasting cultural impression on the Town of Pilot Mountain. Unlike the settlement of Eastern North Carolina, with large plantations, widespread use of slave labor, and the establishment of the Church of England, the Piedmont region, including Pilot Mountain, was inhabited mainly by subsistence farmers, with smaller farms, with few or no

8. Reichel, *The Moravians in North Carolina: An Authentic History*, 200–201.

> slaves. The fertile land provided excellent sites for mills, farms, and especially the growth of tobacco.
>
> —Town of Pilot Mountain (2022)[9]

Having fertile land with good water around a Native American mound is another key sign of the Mound Culture for Pilot Mountain. The Piedmont of North Carolina was very different than the rest of the colonies. There was an isolated quality to it. In some respects that was exactly what they wanted, but they were also treated differently. Most of all, they wanted freedom and to be left alone.

> The Piedmont was isolated from Eastern Carolina by poor and nonexistent roads, by great differences in culture, habit, and custom. They were isolated by the lack of commercial centers and markets for salable merchandise. They were isolated by serious differences in religious belief and practice. They were separated by great differences in attitudes about slavery. They were separated by an absentee government expressing great bias in laws, tax collecting methods and representation in that government. An Eastern planter could pay his tax in goods; cash was demanded of the Piedmonters. The pacifist Quakers and Moravians paid a heavy penalty by not participating in the militia. They were forced to pay quadruple taxes, and during the war even more.
>
> —Fred Hughes (1988)[10]

MORE EARLY SETTLERS ARRIVE AROUND PILOT MOUNTAIN

The Moravians and Quakers were among the first to settle around Pilot Mountain, but there were others.

> The region also offered a variety of religious denominations not seen in the eastern parts of North Carolina, including: Quaker, Lutheran, German Reformed, Moravian, Dunkard, Baptist, Presbyterian, and Methodist. This variety of religious organizations is still evident in Pilot Mountain.
>
> —Town of Pilot Mountain (2022)[11]

---

9. "History," Town of Pilot Mountain.
10. Fred Hughes, *A Map Supplement* (Jamestown, NC: The Custom House, 1988), 24–25.
11. "History," Town of Pilot Mountain.

196 NORTH CAROLINA HISTORICAL COMMISSION

Entry	Amount
May 22nd 10 Indians on their return from Winchester had dinner	6; 8
May 29th had several Indians, dinner	3; 0
June 13th, in the afternoon came 6 Indians on their return had dinner	4; 0
and June 14th breakfast	3; 0
June 21st at noon returned 10 Indians, had dinner	6; 8
and supper	5; 0
June 26th came a company of 100 Indians, had dinner and supper in one	2;10; 0
June 27th 100 had breakfast	2;10; 0
had dinner	3; 6; 8
had supper	2;10; 0
June 28th, they had breakfast and took provision along	2;10; 0
July 2nd, 50 Indians had dinner	1;13; 4
Aug. 21st 60 Indians had dinner and supper together	2; 0; 0
Aug. 22nd they had breakfast and provision upon the road	1;10; 0
Aug. 23rd 3 sick Indians had provisions	2; 0
Oct. 26th 6 Indians had supper	3; 0
Oct. 27th they had a rest day, breakfast	3; 0
dinner	4; 0
supper	3; 0
Total	45; 0; 0

FIGURE 2.5. This is an example of some of the records kept of visits to the Moravians by the Native Americans. Courtesy of Moravian Archives, Winston-Salem, NC.

THE DUTCH FORT

At this time, the area around Pilot Mountain was still Cherokee hunting grounds. There is documentation of them being seen in this area often.

> Occasional detached companies of Cherokee warriors, as also several bodies of Creek and Catawba Indians, passed through the settlement, or encamped near the mill. Receiving plenty to eat, they behaved very well, and gave no cause for complaint. Sometimes there were accompanied by British officers, who paid for them. At other times, coming alone, with a passport of the English government, they were freely received and hospitably entertained (the government of North Carolina afterwards remunerating the Moravians) In consequence, Bethabara became a noted place among Indians, as the 'Dutch Fort, where there are good people and much bread.'[12]

12. Fries, *Records of the Moravians*, 1:196.

In October of 1759, the Cherokee and Creeks declared war against all white people.[13] By that time, the Moravians had built Bethabara and the new village closer to Pilot Mountain, four miles away from their first settlement, which was called Bethania. The Moravians built a fort at Bethabara and many in the area took shelter there. A typhus outbreak in the crowded conditions killed many at the Moravian's so-called "Dutch Fort."[14]

FIGURE 2.6. Recreation of the Moravian Fort from the ground at Bethabara Historic Park, often called the Dutch Fort.

> Almost daily, either Br. Spangenberg or Br Ettwein accompanied by some Brethren, went to Bethania, one going and remaining there, the others returning. 'On one occasion,' Br Ettwein relates (probably in March, 1760), 'when early in the morning tracks of Indians had been observed, the accompanying Brethren were rather fearful, because we generally rode quite slowly, and were talking among themselves how they might make Spangenberg ride faster. When they came to the dense woods, where the most danger was to be apprehended, Spangenberg said: 'You don't know how to ride; let me lead.' Saying which, he set off at full speed, never stopping till they came to Bethania. There Spangenberg remained, whilst he returned to Bethabara, but was treated with less ceremony."'It is not yet safe,' my companions said; 'we must ride as fast as we can; Spangenberg has also done so; and thus, we were racing day after day." It was subsequently proved that this precaution, as well as the orders of Br.

13. Fries, *Records of the Moravians*, 1:48.
14. "North Carolina American Indian Timeline," North Carolina Museum of History: A Smithsonian Affiliate, accessed November 9, 2022, https://www.ncmuseumofhistory.org/american-indian/handouts/timeline.

> Spangenberg to have the church bell rung every morning at dawn of day, was not needless.
>
> Often in the morning the traces of the Indians were found quite near the houses, and it was afterwards ascertained, through some who had been prisoners among the Indians, that one hundred and fifty of their warriors had encamped for nearly six weeks about six miles from Bethania, whilst a smaller camp was only three miles distant, from the Moravians in North Carolina. Several times they were on the point of attacking the Fort of the Dutch, but when they came near they heard the big bell, a sign that they had been discovered. Their design of taking prisoners between the old and new town had been unsuccessful; 'for,' as they expressed it, 'the Dutchers had big, fat horses, and rode like the devil.' Thus, under the kind providence of God, no assault was made upon either of the two settlements; but still a strict watch was kept by day and night, the new burying-ground, which was cleared in December, 1757 (being situated on the top of a very high hill), proving a very convenient place for this purpose.[15]

What the Cherokee heard was just the morning bell that a new day had started, but they misinterpreted that their early morning surprise attack had been discovered. Because of sides the Cherokee took during wartime, their relationship with the Moravians was strained but could be mended. Ultimately the Moravians would renew these relationships after the French and Indian War and the Revolutionary War. One way the Moravians showed their love and respect was by going to Oklahoma to help the Cherokee settle after being removed from their land via the Trail of Tears. The Moravians had to remove themselves from many areas of the planet, so they must have seen the Cherokee as kindred spirits in that regard. The respectful relationship between the Cherokee and the Moravians is just fascinating.

> Except for salt, the upper Piedmont was self-sufficient. Iron was being made by David Allen, on the Yadkin, by 1765... The latest in books was brought down. The Quakers regularly sent new publications and periodicals to southern meetings. They were disseminated or read aloud during meeting. The Quakers, the Presbyterians and the Germans regularly sent their children to Pennsylvania and New Jersey for education and occupational training. Many Quakers went to Pennsylvania, served an indenture, and returned with a valuable trade or occupation.
>
> —Fred Hughes (1988)[16]

15. Fries, *Records of the Moravians,* 1:49–51.
16. Hughes, *Map Supplement,* 25.

FIGURE 2.7. Aerial view via an airplane flown from Mount Airy Airport of the current recreation of the first Moravian settlement in North Carolina. Finding this location from the air under Piedmont Triad International (GSO) restrictions for flight height in the urban sprawl was difficult.

THE TRANSITION OF THE GREAT WARRIOR PATH TO A ROAD

Slowly, the narrow Great Warrior Path was improved, widened, and maintained to become the Great Wagon Road used by the Moravians with Pilot Mountain as the major landmark. Seeing Pilot Mountain from the Blue Ridge Mountains brought such joy to the travelers for decades. In historical documents, it was written as the major landmark to look for on your journey to a new home. One can imagine even the toughest among them getting misty eyed seeing that mountain in the distance after weeks of hard traveling. For decades, the sight of Pilot Mountain was a blessing for these people desperately searching for freedom.

> Mile by mile, the Wagon Road spread further into the Deep South. After Ingles' Ferry was established, travelers by horse and wagon could travel with some assurance down the Appalachians as far as the Yadkin River in North Carolina, though the road grew progressively worse.
>
> Once the Yadkin River was reached, the road branched into several old Indian trails which had developed in earlier days between villages of the Occonneechee, Tuscarora, Catawba, Shawnee, Cherokee and other tribes. The growth of the Moravian settlement of Wachovia after 1753 increased travel from Virginia south to that region. So many settlers were now coming into the Piedmont North Carolina that the county seat of

> Salisbury established thirty miles south of Wachovia, at the juncture of the old Catawba and Cherokee paths.
>
> —Parke Rouse, Jr. (2016)[17]

The settlers came down in Conestoga Wagons with household goods, but those same wagons were also used as freightliners hauling goods. Some of these wagons carried over ten tons of goods. All sorts of animals like sheep, chickens, and pigs would walk right along with the wagon. Not everyone could afford one of the big wagons, so little two-wheel carts were also found along the road.

By 1765, most of the road had been cleared to accommodate horse-drawn vehicles. The gravel of the road deep in the ground will be revealed by ground-penetrating radar for any archaeologist looking for the old road.

> To maintain it, county courts appointed 'overseers' and 'viewers' who were responsible for keeping up segments of the thoroughfare at county expense. To fill the holes and lay new gravel over last year's mud, local farmers were employed in the fall, after they had gathered the crops. 'Road work' remained a source of off-season income for rural Americans for many years thereafter...
>
> Pack horse trains vied with wagons as carriers of the frontier's goods. Each horse in the train was fitted with a pack saddle, which was strapped to its back. Cargo weighing as much as 600 pounds was sometimes carried. A rider on the lead horse led as many as ten or twelve horses in procession, the bridle of each being attached to the saddle of the preceding horse. When staked out to forage at night, pack horses were often belled so they could be followed if they strayed. Many pack-train leaders wrote of the trouble of rounding up a pack train dispersed by storm or Indian attack.
>
> —Parke Rouse, Jr. (1992)[18]

ENTERTAINMENT HOUSES ALONG THE ROAD TO NORTH CAROLINA

Coming down the Great Philadelphia Wagon Road, there were some entertainment houses to stay at, but they were few and far between the further south you got. In the early days of the road, each inn was usually spaced about one day's

17. Park Rouse Jr., *The Great Wagon Road: From Philadelphia to the South* (New York: Dietz Press, 2016), 70. First published 1973 by McGraw-Hill.
18. Rouse, *Great Wagon Road*, 164.

journey apart. A sign on an inn or entertainment house, which were also called ordinaries or public houses, didn't really explain what they offered. Susan P. Schoelwer works for the Connecticut Historical Society (CHS) in Hartford and writes about their museum collection of seventy inn signs.[19] It is estimated there were five thousand signs, according to license records of the time, but only one hundred remain to this day. CHS currently has the largest collection of inn signs in the world dated from 1750 to 1850. These signs combined the artistry of woodworking, painting, and metalworking.

FIGURE 2.8. Looking out the front door of a cabin home in Historic Bethabara Park.

> What is a tavern sign? The term designates a popular category of early American folk art, typically consisting of a wooden signboard painted on both sides with a combination of images and words and equipped with hand-forged, often decorative, iron hardware for hanging. Tavern signs originated from the practical need to identify dwellings licensed to provide entertainment to travelers—entertainment being the period term for essential services, including food, drink, and lodging, as well as feed and stabling for horses...

19. Susan P. Schoelwer, "Tavern Signs Mark Changes in Travel, Innkeeping, and Artistic Practice," Connecticut History, accessed November 9, 2022, https://connecticuthistory.org/tavern-signs-mark-changes-in-travel-innkeeping-and-artistic-practice/.

> Whatever their shape, signboards were painted on both sides and hung for maximum visibility–usually from a crossbar atop a high post or from a horizontal wooden bracket extending outward from the side of a house. Colonial-era signs customarily displayed a single large image above a brief legend, typically the inn holder's initials, a date (usually the year a license was first granted), and sometimes the word 'entertainment.' The words 'inn' or 'tavern' rarely, if ever, appear on extant signs from the 1700s. Surviving imagery includes horses, soldiers, suns, birds, coats of arms, ships, bulls, and saddles; newspaper accounts and other documents record countless other motifs.
>
> —Susan P. Schoelwer (2022)[20]

The tavern signs utilized specific symbols to lure travelers to that particular inn with only a glance. At first, lone inns were about a day's journey apart, but later, competition made marketing necessary for the professional innkeeper. These varied symbols would attract Masonic Freemasons on the road to a distant meeting and farmers bringing items to market or returning with supplies, and were even military symbols for various branches of service. The key to the settlement of the area around Pilot Mountain and beyond was the Great Philadelphia Wagon Road in terms of transportation of people, goods, and services.

> The umbilical cord bringing sustenance to the Piedmont was the Great Wagon Road. It was a two-way road. The cattle drives were welcomed in Philadelphia, and a great variety of goods was available for the return trip. The Quakers and Moravians had regular courier service to Pennsylvania. The roads were passable, with established travel stops. The travelers were safe while on the road. Powerful family ties connected them to Pennsylvania, and they could visit family and friends while there. Cultural and commercial advances were brought to Carolina almost immediately. The cotton gin was invented in 1793. There is no way to determine when they were acquired but wills listed cotton gins: John Hough 1801 (Yadkin), 1803 (Surry), Abner Sharpe 1807 (Iredell) George Mendenhall 1805 (Guilford).
>
> —Fred Hughes (1988)[21]

20. Schoelwer, "*Tavern Signs Mark Changes.*"
21. Hughes, *Map Supplement*, 25.

REVOLUTIONARY WAR IN THE PILOT MOUNTAIN NEIGHBORHOOD

Everyone who lived around Pilot Mountain during the American Revolution found themselves right in the middle of the war. In no uncertain terms, being surrounded by the war meant your village or farm was the resupply area. The cannon and musket fire got closer and closer each day, and soon the army would be on your very doorstep demanding everything you had that wasn't nailed down. Even your fences were used as firewood right in front of your home by the armies of each side to stay warm and cook.

Fred Hughes' view on local history around Pilot Mountain is an interesting one and is informed by his own service to his country at a time of war. Hughes served in the parachute troops in Italy during WWII. Hughes participated in the invasion of France and the Battle of the Bulge. Hughes was awarded a Silver Star, four Purple Hearts, a Bronze Star, the Fouragere and Combat Rifleman, and received a battlefield commission as a Lieutenant.[22] Here is mapmaker and researcher Fred Hughes's non-textbook look at the Revolutionary War in the Piedmont of North Carolina:

> By standards of the twentieth century, the war of the revolution was a very strange war. This was not a single war, with two antagonists, instead the primary war between two, expeditionary English forces and the Continental line of American forces masked and concealed a very serious and bloody civil war. Protagonists of both sides of the civil war had goals established, with definite plans of procedure and method. Both wars had two opposing armies – a total of four armies in the field. Both wars were responsible for producing great injury, great numbers of casualties, tremendous suffering, and great misery. All four armies tried to 'live off the land,' so the average citizen, living in a war zone could expect foragers from all four armies to pay them visits at their homes and farms. There was no army supply in existence then, being supplied from home base. The foragers would demand fodder and grain for the horses of the army, and food and supplies for the troops. Requisitioned were horses, cattle, sheep, goats, fresh meats and cured meats, vegetable, fruits, flour and meal. They did not want the surplus – they wanted all. Not one of the armies objected to leaving people with nothing. Hunger always followed the path of an army. Of course, guns, gunpowder, and lead were requisitioned at every opportunity. Vouchers for goods that had been requisitioned would represent a loss when paid by the government, and the British vouchers would never be paid.
>
> —Fred Hughes (1988)[23]

22. Hughes, *Map Supplement*, endsheet.
23. Hughes, *Map Supplement*, 103.

This would be an example of what Fred Hughes was talking about during the Revolutionary War when Cornwallis occupied Bethania from the Moravian journals:[24]

February 9, 1781. Cornwallis arrived in Bethania with his army and made camp. The houses were filled with officers and their servants while the camp extended for two- and one-half miles. All the food and livestock including horses were seized, nor did they fail to drink their entire stock of wines and whiskeys.[25]

February 9, 1781. From Friedberg we heard a most distressing report that several of the planters in that section, and among others Martin Walk, a communicant Brother, had been arrested by men in English uniform on the charge that they were enemies of the country. In the afternoon we heard that the English army was in Bethania; we also received orders to furnish meal, which was confirmed in the evening by an English dragoon who came here. It rained all day and all night, therefore no service was held except at dusk, when reports from Greenland were read.

February 10, 1781. During a little conference about the troops which are to be expected today, the Brn. Bibighaus and Holder returned from General Greene, with his answer that he could not protect us, as the English must be already in our towns. When they arrived, we were worrying about them. About ten o'clock the British dragoons came, and then the entire army followed in irregular order, continuing until four o'clock in the afternoon. Among them there were many people who to save their lives had placed themselves, their wives and children under the protection of the army. Major Boss announced that Lord Cornwallis would stop for a while with Br. Bagge, which he did soon after, with him being General Lesley, General McCloud, Major England, Governor Martin, and the Commissary's assistant Booth. The Brn. Marshall and Bagge received them, and were courteously treated by the Generals. After a stay of about an hour and a half the Generals took a friendly leave, especially Governor Martin, the others followed them about four o'clock. The people who followed the baggage stole various things at the store and in the houses, for example the Single Brethren lost all their wash, Br. Meyer lost nine head of cattle, Br. Bonn lost £40. The Single Brethren furnished brandy and oxen, and their wagon made two trips taking meal from the mill to the camp near Friedrich Miiller's. There was much work in providing the hungry soldiers with bread and meal; because of this the

24. Surry County Revolution Bicentennial Commission, "List of Historical Events in Surry County," Surry County Digital Heritage, https://www.surrydigitalheritage.org/s/surry-digital-heritage/item/4603.

25. Fries, *Records of the Moravians*, 1:105.

prayer meeting could not be held. Various Brethren were kept in the camp all night because of business; the camp extended from Friedrich Miiller's to Love, about two and a half miles. In the evening some well-behaved German men were in the town; they live below Salisbury, and were going from the army to their homes, as others are doing. The night was peaceful.[26]

Here are some examples of the Moravian journals with some references to nearby Pilot Mountain, a visit by President George Washington, hardships with money during the Revolutionary War, earthquakes, weather, and much more. This gives just a hint of the six thousand pages in the thirteen volumes of the Moravian journals for the eighteenth and nineteenth centuries:

FIGURE 2.9. The hearth for cooking and warmth inside one of the cabin homes at the Historic Bethabara Park.

April 21, 1763. The Brn. Ettwein and Gammern returned toward evening. They had climbed the rock of the Pilot, which is a wonderful creation, reckoned to be more than 200 ft. high, and more than 500 yards around, composed of the best whet-stone sand-stone. From the top one sees the Brushy Mountains and the Blue Mountains and a high range beyond New River, otherwise the land, far and wide, looks like a beautiful plain.[27]

26. Fries, *Records of the Moravians*, 1:271.
27. Fries, *Records of the Moravians*, 1:557–58.

1764. The Wachau is not hilly, but really mountainous, though the mountains are not higher than those on which Herrnhaag is built or Bethlehem [Pa.]. The ridges are so joined together that no matter where I stand it is possible to go to any other part of the land that I wish without crossing a stream, though the path may resemble the moves of a piece in a game of draughts. And as the mountains are all about the same height it is easy to understand that one can get two different profiles of the land. The ridges give an almost straight horizontal line, and that is why the country looks practically level when seen from the Ararat Mountains or from the Pilot [Mountain]. But the other and more correct profile can easily be pictured, especially when one takes a map and traces the hundreds and thousands of valleys, some of them long valleys taking one, two or three, or even seven and eight hours to traverse. The mountains are not high, as already stated, and are generally steep on one side and of gradual slope on the other. On account of the many mountains, and the usual steepness of one side, and the deep and boggy streams, it is difficult to make good, direct roads, or at least it takes an expert to lay them out.[28]

May 18, 1772. This morning a company of gentlemen arrived from Salisbury, some of them being from Charlestown. They visited the Hutberg and our God's Acre, and were pleased with its order. In the afternoon they left, some, including Mr. Martin, the lawyer, on their way to Court, and the others to see the Pilot Mountain.[29]

April of 1776. From this time on hard money disappeared. Four kinds of paper money were in circulation, - North Carolina, South Carolina, Virginia, and Congress money. The last named went best for a few years, as it could be used in all the states; the North Carolina money, on contrary, could hardly be used outside North Carolina, at least without a 7.5 to 25% loss.

Now everyone came to spend his money where things could still be found, and very few would take change. If there was some part due on a bill, they wanted to spend they would say, if in the tavern: 'Give me a dram for it'; in the store: 'Give me some thread, needles, tape, sugar,' or whatever; to the tanner it was: 'Give me a strap, a pair of soles'; to the potter: 'Give me another pipe,' etc. So, the time began when it was a real problem to spend the money that one was obliged to take in. Even among ourselves there was murmuring on the part of those who had to live on their wages, for goods were higher, but the actual value of what was received in the shops was less, and so wages did not suffice for expenses. About this time, we began to lend our surplus paper money, hoping some

28. Adelaide L. Fries, *Records of the Moravians in North Carolina, vol. 2, 1752–1775* (Raleigh: Edwards & Broughton Printing Company, State Printers, 1925), 557–58.
29. Fries, *Records of the Moravians*, 2:734.

time to get it back in good money, but this did not happen in all cases, and a good deal of what was lent was finally lost.[30]

September 15, 1777. The Brn. Meinung and Miksch went to the west side of Wachovia and to our English settlement to survey land for Fidler and Cheaty. Yesterday Br. Fritz held the funeral service for a woman, a sister of Mrs. Banne; about 150 persons were present. He has been invited to preach in the neighborhood of Pilot Mountain. The weather is rainy.[31]

February 17, 1778. Jacob Tanner arrived with one wagon, which had a load of goods for our store. His second wagon had an accident on the way, as it was fording a stream, for a mill-dam above the ford broke, and the water rushed down upon the wagon and team. The teamster cut his horses loose and came out with them, but the wagon remained in the water for twenty-four hours before it could be taken out, and about twenty bushels of salt, which was in barrels, was all melted. In the great fire in Charlestown, on Jan. 11th, we lost eighteen bushels of salt and some other goods.

March 28, 1781. Palm Sunday. Toward evening the Brn. Christoph Reich and Huter came from Salem, planning to visit Pilot Mountain tomorrow.

March 29, 1781. They returned, having accomplished their purpose, and we thanked the Saviour for their safety.[32]

PRESIDENT GEORGE WASHINGTON'S SOUTHERN TOUR IN 1791

May 31, 1791. At the end of this month the congregation in Salem had the pleasure of welcoming the President of the United States, George Washington, who was returning from his tour of the Southern States. We had previously been informed that he would pass through our town on his way to Virginia. Today we received word that he had left Salisbury, thirty-five miles from here, this morning, so the Brn. Marshal, Kohler, and Benzien rode out to meet him. As he approached the town several tunes were played, partly by trumpets and French horns, partly by

30. Adelaide L. Fries, *Records of the Moravians in North Carolina, vol. 3, 1776–1779* (Raleigh: Edwards & Broughton Printing Company, State Printers, 1926), 1030–31.
31. Fries, *Records of the Moravians*, 3:1162.
32. Adelaide L. Fries, *Records of the Moravians in North Carolina, vol. 5, 1784–1792* (Raleigh: The North Carolina Historical Commission, 1941), 2314.

trombones. In his company were only his secretary, Major Jackson, and the necessary servants. As he descended from his coach, he greeted those who stood around in a friendly manner, showing his good will especially to the children who were there. Then he talked on various matters with the Brethren who had accompanied him to the room which had been prepared for him. At first, he said that he was leaving in the morning, but when he heard that the Governor of this State had expressed a wish to wait upon him the next day, he decided to rest here over one day. He sent word to our musicians that he would like music during his evening meal, and it was furnished for him.

June 1, 1791. On the following day, that is on June 1st, the President and Major Jackson, guided by several Brethren, visited workshops, the Choir houses, and other places in our town, and he expressed his approval of them, especially of the waterworks and the service they gave. An address had been prepared, in the name of the Brethren in Wachovia, expressing our dutiful sentiments toward the government of the States, and the President set a time at which he would receive it. In accordance therewith, at two o'clock it was presented to him by several Brethren, and after Br. Marshall had read it, as is customary, and presented it to him, the President in the same manner gave his answer, couched in favorable terms, and both papers are appended to this diary. Six brethren were invited to dine with him, and during the meal music was again furnished.

Many came from the neighborhood, and from our other congregations to see the President, the most notable man in this country; and the President gladly gave them opportunity to gratify their wish.

Toward evening the Governor of this State, Mr. Alexander Martin, arrived from his estate, which is on the Dan River only about forty miles from here. He, with the President and Major Jackson, attended a singstunde in the evening, the singing being interspersed with instrumental selections, and they expressed their pleasure in it. In the evening the wind instruments were heard again, playing sweetly near the tavern. Secretary Jackson inquired concerning our foundation principles, and was much pleased when we presented him with copies of the History of the Unity and the Idea Fidei Fratrum.

At four o'clock in the morning of June 2nd the entire company left, and the Brn. Marshal and Benzien accompanied them to the boundaries of Wachovia.[33]

---

33. Adelaide L. Fries, *Records of the Moravians in North Carolina, vol. 7, 1809–1822* (Raleigh: State Department of Archives and History, 1947), 2324–25.

EARTHQUAKES, WEATHER, AND VISITING PILOT MOUNTAIN

Two decades later, the strong earthquakes began to be felt and noted in the Moravian journals, and newspapers of the time reported them in some detail.

> **November 22, 1811.** This morning at three o'clock we felt an earthquake, which was particularly noticeable in the country around us. No damage was done.[34]
>
> **December 16, 1811.** In the third hour of the morning there was a fairly strong earthquake shock, which was repeated in the eighth hour, but much less severely. [In another journal]... Between two and three o'clock this morning an earthquake was felt in our town and in the neighborhood. Many of our members were awakened, and in the church and parsonage the doors rattled suddenly and waked us.[35]
>
> **December 27, 1811.** In the afternoon there was the first snow of the winter.[36]
>
> **January 11, 1812....** in the morning the thermometer stood at zero.[37]
>
> **January 23, 1812....** this morning about nine o'clock we again felt an earthquake.[38]
>
> **February 7, 1812.** In the morning in the fifth hour, and again in the evening between eleven and twelve o'clock, we felt such severe earthquake shocks that the room-doors in some of the houses were forced open. It rained all day.[39] [In another journal]... This morning about four o'clock we had a very noticeable and lengthy earthquake, which waked most of the Brethren and Sisters from sleep. In the evening between nine and twelve there was another distinct earthquake. We heard later that this extended over a wide area, especially the southern and western part of the United States, and that it caused no little fright in some places.[40]
>
> **January 8, 1817.** In the fifth hour of the morning a slight earthquake was noticed in various places in town, and also in the neighborhood.[41]

---

34. Fries, *Records of the Moravians*, 7:3057.
35. Fries, *Records of the Moravians*, 7:3156.
36. Fries, *Records of the Moravians*, 7:3156.
37. Fries, *Records of the Moravians*, 7:3185.
38. Fries, *Records of the Moravians*, 7:3165.
39. Fries, *Records of the Moravians*, 7:3185.
40. Fries, *Records of the Moravians*, 7:3166.
41. Fries, *Records of the Moravians*, 7:3325.

October 10, 1823. My wife and I visited the widowed Sr. Schneideer, who lives six miles from Bethania, also Salomon Sponhauer, nine miles, and Charles Holder, fifteen miles from here. Holder's wife still belongs to the Bethania congregation. These people live near Pilot Mountain, and we took advantage of the opportunity to climb this remarkable mountain. It rises, solitary and steep, from level land. The top, or crown, consists of one, single, uncut rock, with sides so perpendicular that it can be climbed at only one place, and there with difficulty. On top of this mass of rock there is three-quarters of an acre of fertile soil, in which trees grow. Because of its remarkable formation this Pilot Mountain is surely one of the most notable works of nature in the state. It fills the beholder with wonder, and moves him to praise Him who has wrought so royally. From the top of this mountain one has a very wide view on all sides, and can see several hundred farms, and the winding river Yadkin. On the western horizon the great Blue Ridge is plainly visible.

In the soft light of the moon we walked back to the home of our guide. His wife had wished that I might hold a service at her house that evening, but it was not possible because nearly all the neighbors were busy husking corn. Her eighty-year-old mother lives with them and they took good care of her.

The Bethania gristmill was built, 1783-84, on Muddy Creek, half a mile north of the 2500-acre Bethania tract, near the old road to 'the Hollow,' the section of which Mt. Airy is now the chief town. The agreement of 1783 was signed by six Bethania men, as partners in the mill project. Of these men Philip Transou died in 1793, and apparently Heinrich Sponhauer withdrew. The lease of 1797 names four 'mill-partners,' – George Hauser, Sr., Michael Ranck, Henry Shore, and George Peter Hauser. These four men died before 1823, but the church records do not show who inherited their rights in the mill.[42]

November 6, 1824. Today, as on many earlier Sundays, a company from Bethania went to the Pilot Mountain. The weather was clear and warm.[43]

June 5, 1825. Toward evening a Methodist minister named Schiler preached in our church, at his own request. The man was born in Holstein, had been in Christiansfelf several times. He served as a soldier, and at one time held deistic tenets, but was awakened and now seeks to convince others. At present he is re-painting the house of Mr. Shepherd near Pilot Mountain. He lives in Greensborough.[44]

42. Adelaide L. Fries and Douglas LeTell Rights, *Records of the Moravians in North Carolina, vol. 8, 1823–1837* (Raleigh: State Department of Archives and History, 1954), 3662–63.
43. Fries and Rights, *Records of the Moravians*, 8:3760.
44. Fries and Rights, *Records of the Moravians*, 8:3757.

October 5, 1827. Hard rain about noon prevented me from going to Bethabara.[45]

October 6, 1827. I went to Salem for Sr. Rebecca Bagge, so that she and our family could go to the Pilot.[46]

October 6 and 8, 1827. We made a trip to the Pilot, lodged with the Charles Holders. The weather was fine.[47]

September 24, 1828. Drove in company with Nancy, Br. Daniel Welfare, and Lucinda Bagge to Charles Holder's at the foot of Pilot Mountain in order to hold a service at his request. He is about to move with his children and grandchildren to Indiana.[48]

April 3, 1833. Baptized in Schortown a child of Frederic Krause who lives not far from Pilot (twelve miles north of Bethania), and two children of Jonas and Friedr. Werner in Mother Werner's home.[49]

FIGURE 2.10. A sturdy and well-built Moravian home at Bethabara Historic Park.

45. Fries and Rights, *Records of the Moravians*, 8:3861.
46. Fries and Rights, *Records of the Moravians*, 8:3861.
47. Fries and Rights, *Records of the Moravians*, 8:3861.
48. Fries and Rights, *Records of the Moravians*, 8:3895.
49. Fries and Rights, *Records of the Moravians*, 8:4100.

1836. In the region between Bethabara and the Blue Ridge Mountains, a belt of land extending through Surry, Stokes and Rockingham counties, there are frequent hail storms which cause considerable damage to tobacco and other crops. In the region are Pilot Mountain and the Sauratown Mountains. A study of weather conditions might well be made there.[50]

May 16, 1837. Since there were many visitors here for the coming examination in the Girls Boarding School, Br. Van Vleck preached in English. Among others present was the Cherokee Chief John Ross on his return from Washington, with whom a conference was held.[51]

May 24, 1837. Br. And Sr. Van Vleck returned from a little trip to Pilot Mountain.[52]

July 13, 1841. Today with Mr. Lemly, R. Schweinitz (here on a visit from Pennsylvania) and L. Demuth made a trip to Pilot Mountain. The day was indescribably hot.[53]

FIGURE 2.11. A storm just passing into Stokes County. This is a unique example of the pyramid aspect of Pilot Mountain State Park looking north.

50. Fries and Rights, *Records of the Moravians,* 8:4234.
51. Fries and Rights, *Records of the Moravians,* 8:4264.
52. Fries and Rights, *Records of the Moravians,* 8:4264.
53. Minnie Smith, *Records of the Moravians in North Carolina, vol. 9, 1838–1847* (Raleigh: State Department of Archives and History, 1964), 4625.

June 29, 1847. I visited in the vicinity of Pilot Mountain and there also the wish was expressed that a congregation of the Brethern might be founded.[54]

September 16, 1866. Preached at Stony Ridge School house in Surry County near the Pilot Mountain to a good congregation.[55]

September 19, 1866. Went up on the Pilot Mountain. A Mr. Stevenson preached at the foot of the Pinicle.[56]

April 18, 1873. A cloudy day. At 7 A.M. I with family started for a visit to the Butner families near Pilot Mountain. After a toilsome and fatiguing drive of 16 miles over a very bad road we reach the house of Br. Edward Butner at 1½ o'clock P.M., where we were entertained in the most hospitable manner and tarried for the night.[57]

July 10, 1873. Visited the cascade near and the Hanging Rock on the Sauratown Mountains and came back to the Springs at night.[58]

July 11, 1873. Left the Springs and went on the north side of the Mountains by the Jones Den towards Pilot Mountain and came to Mrs. Gorden's on the Hollow Road, where we staid overnight.[59]

To know what it was like to live in North Carolina for the past 270 years, the Moravian journals are the motherload of detailed information organized by the day and week, plus compiled monthly and yearly wrap-ups. These brief daily descriptions take you back to an entirely different time period. These journals weren't found in an attic or tucked away in a chest in the basement unread. They were written with the intent for current and future generations to read about their past and to learn from it so that past mistakes aren't repeated. These journals were sent out to other Moravians as a way to share news. History repeats itself when information is erased to cover up an unpleasantness. The firm, pacifist stance of these early inhabitants of North Carolina is a wonder and an inspiration to witness. This wasn't a pacifist stance that was shy and hidden behind the drapes within a home or barn. It was facing the armies of both sides during four wars over the course of more than a century. The Moravians were brave souls who walked the walk, both literally and figuratively, in times of trouble for our country with their convictions intact.

54. Smith, *Records of the Moravians*, 9:4979.
55. C. Daniel Crews and Lisa D. Bailey, *Records of the Moravians in North Carolina, vol. 12, 1856–1866* (Raleigh: Division of Archives and History, North Carolina Department of Cultural Resources, 2000), 6658.
56. Crews and Bailey, *Records of the Moravians*, 12:6658.
57. C. Daniel Crews and Lisa D. Bailey, *Records of the Moravians in North Carolina, vol. 13, 1867–1876* (Raleigh: Division of Archives and History, North Carolina Department of Cultural Resources, 2006), 7114.
58. Crews and Bailey, *Records of the Moravians*, 13:7056.
59. Crews and Bailey, *Records of the Moravians*, 13:7067.

FIGURES 2.12 and 2.13. Pilot Mountain from an observation area in Virginia with Winston-Salem in the background on March 21, 2023 (top). First day of spring sunrise from the Blue Ridge Parkway in Virginia on the same date (bottom).

FIGURE 3.1. Stone face on a Pilot Mountain State Park trail.

3

The Frenchman of Pilot Mountain in 1797

FIGURE 3.2. Current view of the Gillam Hotel in 2021 with Pilot Mountain State Park in the background. The hotel and barn were built in 1830 and still stand today.

Starting in 1830, the Gillam Hotel opened for visitors to Pilot Mountain. The visitors and their guide would end the day on Pilot Mountain by taking care of their horse in the barn beside the hotel, cleaning up, changing clothes, and then having a fine dinner with other guests. Fresh picked garden vegetables from behind the hotel and nearby farms would have made this a delicious meal. They had horses, cows, some sheep, and probably chickens at the hotel. If you were one of the lucky ones, you would sit at the table of Andre

Mathieu, who had owned Pilot Mountain for three decades. Maybe that night they had some fish from the Yadkin River for dinner. Then after dinner, out on the Gillam Hotel porch, you might inquire further about a story Mathieu told at dinner. Mathieu's adventure stories of his incredible life would have been so intriguing, but his thick French accent might make a word or two difficult to hear at dinner with many guests talking about their day on the mountain. Out on the porch, the subject at dinner could be brought up again to clarify something or to hear more stories. The view of the Little and Big Pinnacle of Pilot Mountain is stunning from the Gillam Hotel porch to this day with the sun setting behind you at twilight. That porch with that view of the mountain would have been a perfect spot to reflect about what was viewed on Pilot Mountain that day. The building and barn still stand in the same location.

While Mathieu might have told more of his stories out on the porch at twilight on a cool summer evening with the breezes blowing, it was inside the Gillam Hotel that others would gather around the piano singing popular songs of the day. Instruments were laid around the living room for anyone to pick up, after asking permission and if they knew how to play. The hostess of the Mathieu Hotel was Mathieu's daughter, Teresa Hannah Gillam. In 1823, she married Judge William Gillam and they manage the hotel together. Teresa was also a gifted musician and artist. Special nights were when a few choir-trained voices visited to harmonize together.

It was reported on August 13, 1851, that the Gillam Hotel had 253 visitors just since July 1 of that year.[1] Clearly, the publicity campaign during the decade before this report worked:

> At the foot of the mountain is an excellent public house, kept by Wm. Gillam, Esq., whose amiable and tasty lady makes everybody comfortable and happy who favor her with their company. The charges are remarkably low, and the fare is of the most substantial, and even delicate kind. The house is large and would accommodate 30 to 40 persons.
>
> Some 400 yards from the house, is an excellent mineral spring. There is a fine road to the house, and I can truly say that no one can visit a scene more beautifully grand than the Pilot mountain, or stop at a house where they will fare better than at Mr. Gillam's.
>
> As a place of resort for health, I know of no place superior to it.
>
> —[signed] M. B. (1850)[2]

While many ads were published across North Carolina for the Gillam Hotel, this one in 1855 was the first found to explain the prices at the Gillam Hotel. Almost all the stagecoach ads in this area mentioned the price to go from

1. Staff writer, *The Weekly Standard*, August 13, 1851.
2. M. B.,"A Day at Pilot Mountain, NC," *The North Carolinian*, October 4, 1850.

Greensboro to Pilot Mountain specifically. The railroad wouldn't arrive until over three decades later, in 1888, so walking, horse, buggy, or stagecoach were the only options for travel to Pilot Mountain.

> Pilot Mountain: The public are informed that the ascension to the top of the Pinnacle has been made safe and easy, by iron hand-rails and steps where it was necessary, and a good carriage road much farther up the mountain than formerly.
>
> Travelers passing across the Blue Ridge can visit this place and only lose 4 miles on the route. Prices of Board.
>
> | | |
> |---|---|
> | Man and horse, per day | $1.50 |
> | Man, per week, | 4.00 |
> | Man and horse, per week | 7.50 |
> | Single person, per month | 12.50 |
>
> August 24, 1855
>
> —Wm. Gillam[3]

People would be arriving at the hotel at the same time people were coming back from their all-day excursions on Pilot Mountain. There was also a special sunset trip with a return in the dark by the light of the moon:

> The mountain with upwards of ten thousand acres of land surrounding it, is now held by Wm. Gilliam, Esq. and family, formerly of Columbus, South Carolina, who keeps an excellent Hotel at the foot of the mountain on the south side. Every thing about his establishment conspires to make the weary pilgrim to this shrine of nature feel comfortable, and his interesting lady and family do all they can to contribute to your enjoyment and to make you at home.
>
> When I arrived late in the afternoon in front of the house immerging from the deep vales and high hills nearby and slaking my thirst at a mineral Sulphur Spring close by the house, my ears were saluted with the music of the Piano accompanied by vocal singing by the eldest Miss Gilliam who is a most admirable performer on that instrument. I saw on entering the parlor in addition to the Piano, the Harp and Accordion, on both of which instruments she is an excellent performer.
>
> —[N. J. P.] (1847)[4]

3. Ad, *The People's Press*, August 31, 1855.
4. N. J. P. "The Pilot Mountain," *The Weekly Standard*, October 27, 1847.

FIGURE 3.3. This is the old Gillam Hotel, built in 1830, as it looks today in 2022. This location could be a wonderful spot for a business excursion that would recreate "going to the mountain" in the 1800s by horseback.

RESEARCH GOLD WAS STRUCK ONLINE FOR ANDRE MATHIEU!

There was one particularly dreary Friday in 2019 when, while looking for something else, I put Andre Mathieu's name into a newspaper search engine and obtained thirty-four returns from all over the eastern United States, including four obituaries unseen up until then. It was like striking research gold that day, so I spent most of the rest of the day obtaining the reports, dates, and sources. The next day, and ever since, there were only three or four returns for that same search, which reveal very little about Mathieu's life. It is as if for a twelve-hour period, the search engine worked as it should but didn't before or ever since. I have mentioned this to librarians across the state and they always nod their heads in agreement and say this isn't unusual. If one is doing genealogy research on this man, the footnotes in this book are invaluable. Finding information about Mathieu is like looking for a needle in a haystack. Clearly, having the world at your fingertips, in a technological sense, for research comes at the price of algorithms that feed the same story to you in an unbalanced way over and over.

Was it just the luck of persistent checking, finding all these items of interest on Mathieu, or did the algorithms just happen to go down for a short time period that day?

THE FASCINATING LIFE OF THE FRENCHMAN OF PILOT MOUNTAIN

Mathieu was born in Checy, France, circa 1761 to a woman named Jeanne Mathieu.[5] A most colorful character, Andre Mathieu joined the French military

5. Mathieu folder, Edith M. Clark History Collection, Edith M. Clark History Room, Rowan Public Library, 201 West Fisher Street Salisbury, NC, 28144; Lelia Graham Marsh, "Letter to the Editor," *Winston-Salem Journal*, December 22, 1966.

FIGURE 3.4. View from the Antigua fort in the West Indies toward where the naval battle called the Battle of the Saintes happened, which owner Andre Mathieu was involved with during his military career. This image was taken by the author.

under either General Lafayette or General Rochambeau (Jean Baptiste Donatien de Vimcai, comte (Count) de Rochambeau) as a private in his teens (circa 1775–77).[6] It isn't clear which general he served first, but it is noted he ran away from home at a young age to volunteer as a private in the Army or Navy.[7] He was present when the combined forces of General George Washington, General Lafayette, and General Rochambeau surrounded and forced Lord Cornwallis to surrender at Yorktown October 19, 1781. Mathieu's hand was wounded by a saber there according to records in the Edith Clark History Collection in Salisbury Library in Salisbury, North Carolina. That wound would bother him the rest of his life.[8]

Several years later, most likely after returning to France, Mathieu served in the French Navy under the French Admiral Count de Grasse.[9] His first naval battle was a famous one called the Battle of the Saintes in the West Indies, near Guadeloupe, which is an island just south of Antigua. This unusual naval battle was fought from April 9 to April 12 in 1782. The French lost that battle. Admiral Count DeGrasse's Flagship was forced to surrender and they were taken prisoner. The navel term "breaking the line" was born during this engagement of hundreds of ships. The direction of the wind played a huge role in this battle.[10]

6. Federal Pension Application: Pre-Civil War, 1775–1783, File NAFT 85A, National Archives.
7. Mathieu folder, 1966.
8. James Brawley, *Footnotes to History: Life Led from France to Salisbury* (n.d.), Edith M. Clark History Collection, Edith M. Clark History Room, Rowan Public Library, 201 West Fisher Street Salisbury, NC, 28144
9. "Andre Mathieu Obituary," *The Pittsburgh Gazette*, December 18, 1857.
10. Donald Sommerville, "Battle of the Saintes: West Indies [1782]," *Encyclopedia Britannica*, https://www.britannica.com/event/Battle-of-the-Saintes.

Losses: British, no ships, 1,000 dead or wounded men; French, 4 ships captured, 1 ship destroyed, 5,000 dead, wounded, or captured men.[11]

There is some mystery about what happened next, but it seems Mathieu must have gone back to France again during the French Revolution. Mathieu served at Toulon in France. In Mathieu's later years, he would remark he saw Napoleon Bonaparte, who was then an artillery captain in Toulon at that time.[12]

It was also mentioned in a newspaper article titled "Original Sketches: Notes on the Way" that Mathieu had dinner with General George Washington in Boston after the Revolutionary War. Since they were both Masonic Freemasons, it might have happened at a meeting at the Masonic Lodge or Inn. Maybe it was to repay some kindness Mathieu had shown to General George Washington during the war. To become a Masonic Freemason, you must self-select to join the secret society, as they do not recruit members and there is a solemn secret oath to take. You don't know the oath until you are taking it.

In his obituary, it was stated that Mathieu was a witness to the French Revolutionists Murat, Danton, and Robespierre's Saturnalia, which might have made him one of the key witnesses in the French Revolution.[13] At any rate, the man got around to many interesting places with many interesting people during a monumental era of change when travel was far more difficult than it is today.

ANDRE MATHIEU GOES INTO BUSINESS IN NYC

At some point in the late 1780s or early 1790s, Andre Mathieu finally ended his military career and returned to live in New York City. Mathieu crossed the Atlantic back and forth several times in the first three decades of his life, which was very unusual in that time period. Usually, it was a one-way ocean voyage and a vow to never return after tasting freedom on this side of the Atlantic. Plus, there was the expense of a trip like that across the ocean to consider.

In 1794, he opened and ran a French coffeehouse at 190 Greenwich Street in New York City. In 1795, he was successful enough to turn the coffeehouse into a tavern and boarding house at the same location. The boarding house moved to 22 John Street in 1797 and then later was called the Mathieu Hotel, which ended up at 31 Nassau Street in 1802.[14]

In just thirty-three years, Mathieu had lived in France, traveled to the colonies, fought in at least two revolutionary armies, participated in a famous naval battle, was possibly held prisoner, traveled between the colonies to France at least twice, and survived with only the mention of a saber wound to his hand from the

11. Sommerville, "Battle of the Saintes."
12. "Andre Mathieu Obituary," *Pittsburgh Gazette*, Friday, December 18, 1857.
13. "Andre Mathieu Obituary," *Pittsburgh Gazette*, Friday, December 18, 1857.
14. "Sixth Regiment," *The Evening Post*, New York, NY, May 5, 1802; New York City Common Council Minutes, November 5, 1805, 625; Directory search for Mathieu, André, New York Public Library.

Yorktown battle (though it did plague him the rest of his life). Now he had become a successful tavern and boarding house owner. What would be next?

Mathieu received a letter from France on August 10, 1795, from his mother, Jeanne Mathieu, that congratulated him upon his marriage, gave news on the French Revolution, and news of his family in France.[15] From this, we can conclude he married his first wife who was a widow, Ms. Hannah Davis Van Deusen, in late 1794 or early 1795, given the time it took to carry letters to Europe and back by ship. Mathieu had three children with his first wife. A miniature of his first wife and one of their children was cherished in the family for generations (whereabouts currently unknown).[16]

It was the tavern on John Street (owned jointly with John Byrne) that acquired Pilot Mountain in 1797 before moving to Nassau Street to become the Mathieu Hotel. It is doubtful that Mathieu or his business partner, John Byrne, would see Pilot Mountain for many years. It was difficult in those days going back and forth from New York to North Carolina on the Great Philadelphia Wagon Road. From the deeds, it looks like lawyers from New York and Philadelphia took care of the deed for Pilot Mountain with all the details that involved. Taxes and fees circled, going hand in hand with every deed written at that time.

For the most part, newspaper stories about Pilot Mountain ownership start with Mathieu as the first owner in the late 1700s and early 1800s with no mention of English royalty owning the land decades before that, nor the Native Americans' shared possession into prehistoric times. Earlier, the Native Americans used the vast area as hunting grounds, for sending smoke signals, and creating fire beacons. They also used various areas of the mountain for vision quests, burials, and ceremonies, but formal ownership of land was foreign to them.

However, the Creek Nation in Oklahoma in modern times had an astounding modern victory in terms of millions of acres of tribal land there (that includes Tulsa, OK). The McGirt decision is having far reaching affects.[17] This U.S. Supreme Court decision could impact tribal land for the Cherokee, Choctaw, Chickasaw, and Seminole nations, who were also forced off their land from 1830 to 1838.[18] While the Cherokee Trail of Tears is the most famous, all five Native American nations were involved. There is a 372-page book on the history of land ownership in North Carolina, and one wonders how far reaching the McGirt decision could be.[19] Is it too early to tell if this could impact ownership of Pilot Mountain State Park? Could this make it tribal land again in terms of ownership and jurisdiction with all the complex details from treaties to sort out?

In the colonial records, there are accounts of petitions for land earlier than when Mathieu and Byrne took possession, but not deeds. These accounts stated:

---

15. Mathieu folder, 1966.
16. Mathieu folder, 1966.
17. McGirt v. Oklahoma, 591 U.S., 140 S. Ct. 2452 207 L. Ed. 2n 985 (2020), https://www.supremecourt.gov/opinions/19pdf/18-9526_9okb.pdf.
18. Annie Gowen and Robert Barnes,"Oklahoma Reels after Supreme Court Ruling on Indian Tribes," *Anchorage Daily News*, July 24, 2021, https://www.adn.com/nation-world/2021/07/24/oklahoma-reels-after-supreme-court-ruling-on-indian-tribes/.
19. Kenneth B. Pomeroy, *North Carolina Lands: Ownership, Use, and Management of Forest and Related Lands*, (The American Forestry Association: Washington, DC, 1964).

"William Shepperd enters 10 acres of land in Surry County on top of Pilate Mountain including the pinnacle on said mountain. July 31, 1779 (warrant granted),"[20] "Jethro Sumner enters 50 acres of land in Surry County on the top of Pilate Mountain including the Nob by the name of general Pilate Mountain (no date given),"[21] and according to correspondence with Pearle Beasley on November 28, 1987, with the spelling on the legal documents that way. Shepperd and Sumner are also mentioned in some newspaper accounts often as a "Letter to the editor" to correct something published about Mathieu or the Frenchman being the first owner. Shepperd clearly was granted possession of the 10 acres but it appears that Sumner wasn't. However, Sumner was granted 640 acres on the northwest side.[22]

At one point in 1796, there were three owners of Pilot Mountain in just six months. The deed was passed around like a hot potato before landing in the lap of the co-owned Mathieu Hotel in New York City. There were thirty-six tracks of land noted, including Pilot Mountain, in the deed that NC Governor Samuel Ashe granted to Martin Armstrong (through a power of attorney Thomas Adam Word dated February 4, 1796). Armstrong then gave the deed to Esq. James Stewart of Tyrone, Cumberland County in Pennsylvania on February 12, 1796. That was then let to William Lewis, a gentleman of Virginia on August 16, 1796.[23] It was after this series of deed transfers that Mathieu and Byrne in New York City obtained the Pilot Mountain deed that totaled 9,231 acres.[24] Was gambling for the deed involved in this quick succession of owners before a card game determined the owner for the rest of the next century and beyond?

TRANSPORTATION IN THE EARLY NINETEENTH CENTURY

New York City to Pilot Mountain was an at least six-to-eight-week difficult journey where crossing every river was an ordeal. As innkeepers of the Mathieu Hotel for so many years, they must have heard firsthand accounts from guests of the oft-used Great Philadelphia Wagon Road going back and forth by horse, stagecoach, or wagon. That dreaded deep dark forest of North Carolina with packs of wild animals, and rumored to have giant humans, also might have been brought up by travelers. Travel on that main road at the turn of the century would be among hundreds of head of cattle or thousands of pigs being herded at the same time in both directions. There were inns to stay overnight, but the further south you got, the poorer the conditions you would find.

"If this is coffee, please bring me some tea; but if this is tea, please bring me some coffee," was standard joke attributed to a refined Englishman on the Great

20. Mathieu folder, 1966.
21. Ruth Minick, letter to Carole B. Sperry, Surry County Historical Society North Carolina, November 28, 1987; Ruth Minick, "William Shepperd First Owned Mountain," *The Mount Airy News*, November 17, 1991, 5C.
22. Minick, letter to Sperry, 1987.
23. Minick, letter to Sperry, 1987.
24. Mathieu folder, 1966.

Philadelphia Wagon Road at the time. The quote was also attributed to many others, including President Abraham Lincoln. It is a great quote about a time lost to the fog of history. There is a blog online that references the quote with fourteen distinctly different footnotes as to its origin.[25]

WHAT BEING A MASONIC FREEMASON ENTAILED IN THE NINETEENTH CENTURY

W. B. "Bill" Hosler, in a bicentennial blog of the Masonic Grand Lodge of Indiana (formed in 1818), wrote about travel to Masonic Freemason meetings. Andre Mathieu was also a Masonic Freemason and this is a colorful description of travel to meetings. Hosler includes where meetings were held and how they were set-up, which is quite forthcoming for the secret society.

> Once you understand symbolic things, you, too, will see symbols everywhere.
>
> —Joseph Campbell (1983)[26]

Shall we take a peek of what it was like to be a Masonic Freemason some 250 years ago?

While Hosler is speaking of travel in Indiana and Ohio, this sounds just like what people experienced in North Carolina during that same time in the back-country around Pilot Mountain. Hosler continues not only about the travel but what the meeting was like, plus where they were held in those early days. We have his permission to publish this here, which is appreciated:

> Back in that day, there were very few choices of transportation by which you could travel across a newly formed state; you could travel by riverboat down the Ohio river (if your town was near the river and you could afford the fare), you could travel via horseback (down rough paths which could hardly qualify as a trail), and if you didn't own a horse, you walked.
>
> No matter which mode of transportation you chose, the journey was guaranteed to be uncomfortable; muddy trails, snow and high winds combined to make the trip difficult. Depending on where you lived, your destination may be several days (or even weeks) away. In the winter you

25. Garson O'Toole, "If What You Gave Me Last Was Tea, I Want Coffee. If It Was Coffee, I Want Tea," quoted in *The Madison Courier*, (Madison, Indiana) January, 1840, https://quoteinvestigator.com/2015/11/11/coffee-tea/.

26. Diane K. Osbon, "Reflections on the Art of Living: A Joseph Campbell Companion," quoted in a month-long lecture series at Esalen Institute, (California: HarperPerennial, 1991).

slept on the cold ground, shivering under a blanket near a fire, eating what meager provisions you brought. During the summer, you endured the heat of the day and hordes of insects. You might encounter highwaymen who would think nothing of robbing you of all your possessions and leaving you for dead in the wilderness. You always ran the risk of a wild animal who might see you as a threat to his domain and an easy dinner. There was no 911 or Auto Club to come to your rescue if you were in trouble. You were on your own.

No matter how you traveled, when you arrived at your destination the accommodations were scant at best. Most of these men would stay in the home of another Freemason or in a local inn. Tavern owners usually offered beds above their establishments; you paid to spend the night and you shared that bed with all of the other travelers. You ate what the tavern served that day. Once your business was complete, you began your return trip home facing the same dangers and discomforts as before.

These men so believed in the Craft that they were willing to endure all of these hardships and dangers, not to mention the days of being away from their families and livelihoods, all in order to help advance the Craft.

Traveling wasn't the only hardships our forefathers had to endure. We are taught that Masons originally met in high hills or low vales, which later became the upstairs loft spaces of inns and taverns, accessible only by climbing a ladder. I have heard stories of lodges meeting in caves, in barns, sometimes even in the home of one of the brothers. One thing is for certain, most of these spaces were not ideal for a lodge meeting. Before a lodge was opened, the Masons had to get to work setting up the room, moving the chairs into position, laying out the jewels and the aprons. If degree work was to happen that night, a brother would draw out the tracing boards on the floor which had to be mopped up after the lodge was closed. Once the lodge was closed, the furniture in the space had to be moved back to its normal position. These buildings were drafty, cold and uncomfortable to occupy.

Over the last century, we Freemasons have become accustomed to meeting in beautiful lodge buildings. Sometimes these edifices were marble palaces in the center of a big city, other times a modest room above a storefront in a small town. These buildings all have modern plumbing, are heated for the winter and sometimes even air conditioned for a pleasant climate in the summertime. Custom furniture was commissioned and purchased which never had to be moved. Beautiful carpets lay on the floor beneath the feet of the Brethren.

It's difficult today for us to imagine the travel involved and the meeting places that our forefathers used to spread the light of Masonry.

Sometimes just the events surrounding these men's lives added even more issues for these men to bear.

Most of us know that throughout the last few centuries many Grand Lodges issued emergent charters to men in order to meet during war time. The daily lives of these men are the hardest thing for me in my comfortable modern life to fathom.

These men would march for hours a day, usually with little sleep and even less food, to a battlefield where they had to make camp, and then risk their lives on the field of battle. Once the battle was over, if they survived the conflict and weren't too badly injured, the men would erect a tent, get out the trunk of Masonic regalia from a wagon, and open a lodge. Sometimes lodge officers had to be continually re-elected, not because the brother quit the lodge, but because the man had been killed on the field of battle. Think of it: these men were hungry, exhausted, and trying to forget the horrors of war they had witnessed that day, but they still thought enough of their obligation to continue to meet, just like they would have back in their homes.

Throughout history, our Masonic forefathers endured hardships of all kinds just to practice what today we take for granted. From rough travel, to bad living and meeting accommodations, to actually risking their lives on battlefields or being tortured in a prison camp for their belief in Freemasonry.

Sadly, not all of these hardships are in our past. Many continue to this day. I recently had the distinct honor to speak with a Brother who asked that I keep his name and his home secret, not because he was worried about his own safety (he escaped and is now in a free country) but because the Brethren of his lodge in his home country are still in peril. This Brother belongs to a lodge in the Middle East. His government has declared Freemasonry illegal. If the location of his lodge is discovered by their local government, his Brothers will be arrested, placed in prison, and after they are tortured into confessing crimes against the state, they would be executed. In spite of the risk, they still meet on a regular basis, in a secret lodge room. They meet and discuss Freemasonry and how it helps them in their lives. While meeting, they keep an eye on each other and if a member or his family needs Masonic charity they will quietly arrange it. They don't allow a dictatorship or the threat of death to stop their belief in the obligation they took.

—W. B. "Bill" Hosler (2017)[27]

27. W. B. "Bill" Hosler, "Dangerous Travels," Midnight Freemason, February 2017, http://www.midnightfreemasons.org/2017/02/dangerous-travels.html.

VARIOUS BUSINESSES OF ANDRE MATHIEU

The boarding house in New York City moved in 1797 to 22/3 John Street right in the heart of New York City's business district.[28] In 1801, they moved to Nassau Street under "doing business as (DBA)" the Mathieu Hotel,[29] which was mentioned as a high-end hotel in the newspapers of the time.[30] This is not surprising as he must have made many important contacts during his military career. The Nassau Street location nestled his hotel between the future Wall Street to the southeast and, in modern times, World Trade Center property to the northwest. The Pilot Mountain deed would be considered a valued asset of the Mathieu Hotel at this point. Even back in the early 1800s, this area of New York City was known as the financial district but included a mix of residential homes at that time. The Mathieu Hotel was used for military gatherings plus the hotel was known far and wide for its "turtle soup" sold by the pound.

> Turtle may be had every day at Mathieu Hotel, no. 31 Nassau-street. - Likewise, Gravy Soups. Families may be supplied by sending at 11 o'clock-Dinners, Suppers and Breakfasts provided at a short notice, in the best manner. Very comfortable Boarding and Lodging. April 5.
>
> —Ad in The Evening Post (1802)[31]

> Mr. Wortman presents his respects to the newly appointed Militia Officers, will attend at Mathieu's Hotel, No. 31 Nassau-street, on Wednesday evening next, at 7 o'clock, to deliver their Commissions, and administer the oaths of office prescribed by law. Monday, April 3.
>
> —Ad in The Evening Post (1802)[32]

> Andre Mathieu, takes this opportunity to present his best acknowledgments to those who have heretofore done him the honor to patronage his exertions, and assure them that nothing on his part shall ever be wanted to afford them the best accommodations that can be reasonably expected from such an establishment.

---

28. New York Council Minutes, 625.
29. Directory search.
30. "Tuttle's Hotel," *The Evening Post*, New York, NY, May 14, 1804.
31. "Turtle Soup," *The Evening Post*, May 22, 1802; "Turtle Soup," *The Evening Post*, October 4, 1802.
32. "Mr. Wortman Presents," *The Evening Post*, May 5, 1802.

> Private parties may be accommodated at the shortest notice, with dinner, suppers &c. November 7.
>
> —Ad in The Evening Post (1803)[33]

To the Mathieu family in New York, a daughter, Anna Therese Mathieu, was born on December 15, 1799. Teresa Hannah Mathieu was born in 1800, and a son, Andre Mathieu, was born on February 24, 1802. The children were baptized at the Reformed Protestant Church in New York City. Sometimes it is mentioned it was two sons and a daughter. Mathieu's life was not all adventure and joy. Tragically, his wife, and it is believed also his daughter, fell victim to yellow fever the summer of 1803.[34] His was not the only family affected that year. It was too common during this terrible plague for people to come down with yellow fever one evening and be found dead the next morning.

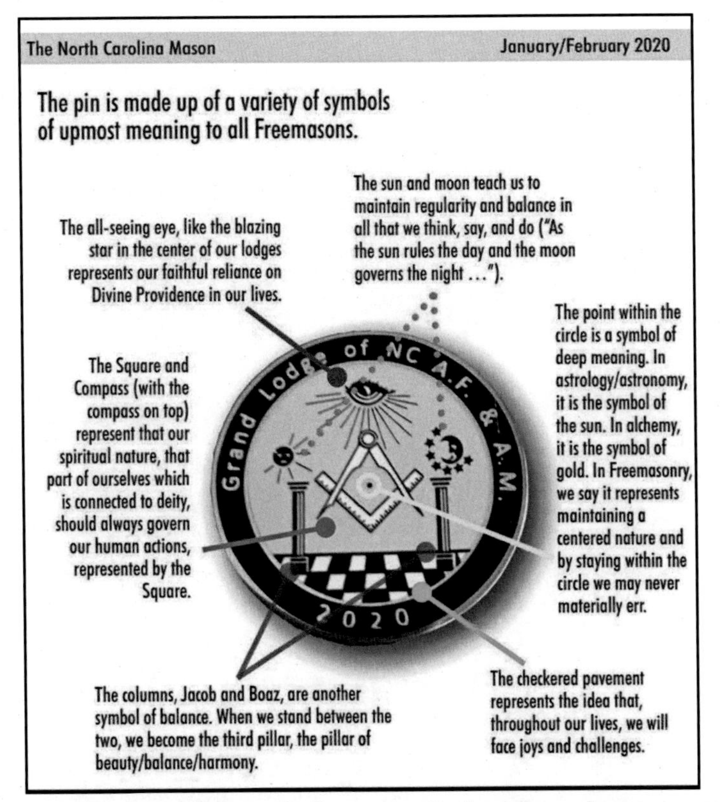

FIGURE 3.5. As Joseph Campbell was quoted earlier, when you see these symbols, then you will see them everywhere. Courtesy of Beth Grace, editor, NC Mason, and made possible by Brother John Pea of The North Carolina Mason.

33. "Andre Mathieu," *The Evening Post*, November 12, 1803.

34. "Died This Morning, Mrs. Hannah Mathieu," *The Evening Post*, September 2, 1803.

DEED HISTORY OF OBTAINING PILOT MOUNTAIN

Dozens of newspaper accounts for the past 200 years talk about Andre Mathieu buying Pilot Mountain and being the first owner, but that isn't the whole truth. There were at least four, maybe five, owners before Mathieu. With lost records of the American Revolution, there may be many more. In addition, was the Pilot Mountain land part of earlier treaties with the Native Americans that could impact modern society?

As already noted, three of the owners had the deed for a short period of time in 1796. The key exchange on October 13, 1797, was when William Lewis, known now as a gentleman of Virginia "sojourning in Philadelphia" (through his attorney John Evans, a gentleman formerly of Tennessee but now of New York City) let the property to Mathieu and Byrne. The deed was a tenants-in-common one, for $1817.75, witnessed by G. Workman and David Cannon and proved December 6, 1797. This is according to the Edith Clark History Collection in the Salisbury Library in North Carolina and the actual deed.[35]

In the written deed, copied from the original, between Lewis, Mathieu, and Bryne, it is described as an "indenture" deed that was exchanged for money or goods that Lewis needed (see fig. 3.6). Much later, in 1915, court documents stated that all of the thirty-six tracts of land from the October 13, 1797, deed totaled 9,231 acres.[36] We will take that court assessment to be a fact. However, the acreage is sometimes reported in newspapers from 30,000 to as high as 300,000 acres, clearly in error, for over two hundred years.

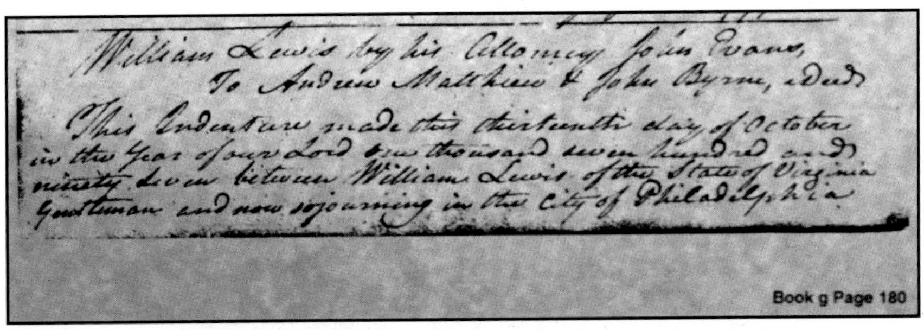
William Lewis by his Attorney John Evans,
To Andrew Matthieu & John Byrne, Deed
This Indenture made this thirteenth day of October in the Year of our Lord one thousand seven hundred and ninety seven between William Lewis of the State of Virginia Gentleman and now sojourning in the City of Philadelphia

Book g Page 180

FIGURE 3.6. This image is the beginning of a long document that informs the reader that it is an indenture deed.

THE LITTLE-KNOWN CARD GAME FOR PILOT MOUNTAIN OWNERSHIP

At some point after his wife's death, Andre Mathieu dissolved his "Mathieu Hotel" business and the assets were divided in an agreeable manner with John Byrne. This was a time when life in New York City was evolving, and a criminal incident might have turned Mathieu against big city life altogether. Mathieu

35. Surry County Deed Office, Dobson, NC, Book G., 180–189.
36. Whitfield v. Boyd, 158 N.C. 451, https://cite.case.law/nc/158/451/.

purchased several small ads during his ownership of the Mathieu Hotel, often advertising the green turtle soup by the pound at the hotel or something about a military gathering, as pointed out on the previous pages.

However, the largest ad he purchased was one for information to capture a thief who had stolen a significant amount of silver from the Mathieu Hotel. It was a long, itemized list of each piece of silver and the name and description of the suspect. That incident, and maybe other unreported thefts, might have influenced his decision to sell the hotel, divide the assets with his partner, and eventually leave New York City.

> Fifty Dollars Reward: Stolen yesterday morning about 7 o'clock from Mathieu's Hotel, in Nassau street, the following articles: - 2 Silver Coffee Pots, one marked Z. on the beak, the other no particular mark; Two dozen Silver Table Spoons, marked A.M.; Eleven Silver Tea Spoons, four marked M.; One Silver Soup Ladle; One Pair Silver Candlesticks; and One Pair Silver Sugar Tongs, marked E. R. – They were stolen by an Irish girl named Ellinor M Kenny, an old offender, with light brown hair, and a scar on the left side of her neck with the King's Evil. She was recommended to the subscriber as a servant maid by the wife of a carman living in Barley-street, whose name he does not now recollect. She was seen yesterday morning to come out of John-street, with some of the above described goods, cross Broadway and go down Dey Street. As it is probable she is secreted in some part of the city, a reward of 20 dollars will be given to any person who will give information where she may be found, or 50 dollars for the Goods and the Thief. Silversmiths are requested to be on the lookout, and if any of the above plate is offered for sale, to secure the person offering it and give information to –
> A. Mathieu, 9 Nassau Street.
>
> —Ad in The Evening Post (1804)[37]

The very last asset to divide for the Mathieu Hotel when it was dissolved, according to two newspaper accounts, was the Pilot Mountain deed. This is where the card game comes into play with just Pilot Mountain in the pot for the winner.[38] They came to that forfeited Pilot Mountain "indenture" deed for non-payment from Lewis and didn't know what to do with it. Surely the division of the mountain hundreds of miles away in the backcountry of North Carolina was discussed at length. The distance from the mountain was likely a key factor. The division of Pilot Mountain would have involved a new survey, lawyers, a new deed, taxes, fees, a great deal of time, and so much more because of the distance

37. Andre Mathieu, "Fifty Dollars Reward," *The Evening Post*, New York, NY, March 19, 1804.

38. Staff writer, "Swepson, Clark and Dunham," *The Raleigh Sentinel*, November 30, 1874; N. J. P., "Pilot Mountain."

involved. A simple card game similar to Euchre, familiar to both, was an efficient way to settle the matter and was agreed upon by both partners.

This, along with the difficulties and expense to legally divide it, must have contributed to it being valued so little as to give it up in a card game! One wonders if neither Mathieu nor Byrne really wanted the Pilot Mountain deed. In their minds at that moment, did the winner really lose and the loser really win? Neither would know what Pilot Mountain looked like or its prehistorical sacred significance to the Native Americans.

> The Pilot Mountain was won at a game of Seven Up by old Mr. Matthews [a.k.a. Andre Mathieu], of Salisbury. When he and his partner closed business in New York,... they had divided the notes and accounts, the question was how to divide the mountain. It could not be done without spoiling its beauty and symmetry; besides it could take a year or more to divide it. The partners agreed, and did play a game of six cards Seven Up for the Pilot Mountain, and Matthews [Mathieu] won.
>
> —Staff Writer (1874)[39]

Since Seven Up, a popular game of the French Navy, usually lasts only twenty minutes, the card game quickly decided who got the deed to Pilot Mountain. The winner turned out to be none other than our world-traveling Frenchman, Andre Mathieu. As late as 1870, people would jokingly play the card game Seven Up on the mountain near the sheep house (circa 1815) on the Pilot Knob Trail.[40]

However, Mathieu and his family did not leave New York City for a little while after the card game was won. Members of the extended Mathieu family owned the mountain until 1915, at which time it was sold at auction.

> This property was entered as vacant land by a company some fifty years ago, and afterwards sold to a mercantile house in the city of N. York, of which Mrs. Gilliam's father, a Frenchman by birth, was one of the members and to whom it was allotted in the division of their estate. Much of the land is first rate for tobacco, corn, and wheat, and but for its remoteness from market would be very valuable. On the North side, the mountain is covered with snow nearly the entire winter, and until the spring is far advanced.
>
> It is contemplated to establish a Female Boar[d]ing School at Mr. Gilliam's, which will be very desirable location for health, beauty of scenery, and cheapness of living. I witnessed a painting by Miss Gilliam of

39. N. J. P., "Pilot Mountain".
40. J. A. S., "From the Old North State: A Visit to Pilot Mountain," *Carolina Watchman*, June 10, 1870.

> the mountain which was very exact and beautiful. My sheet is full, and I must defer my account of the Iron Works and the County of Surry to a subsequent number.
>
> —[signed] N. J. P. (1847)[41]

FIGURE 3.7. Can you imagine the first time Andre Mathieu saw Pilot Mountain after owning it for years? How much of the Big Pinnacle was hidden from view by the virgin timber, which Mathieu spoke of proudly in 1800s newspaper reports?

On May 14, 1804, The Evening Post reported that Johnson Tuttle had taken over the Mathieu Hotel in the "business part of town" and renamed it Tuttle's Hotel at 31 Nassau Street.[42] By August of 1805, Tuttle was trying to rent or sell the Tuttle Hotel from New Jersey.[43]

While Mathieu did leave the hotel business, he didn't leave New York City right away. Mathieu made application for a ten-acre vineyard in 1804, according to the Minutes of the Common Council, but this might have been near the time when he closed all of his businesses or at least within a few years.[44] An ad on June 2, 1804, for the new wine shop said Mathieu had "laid in a stock of choice Liquors, Wines & C[heese] and where individuals and private parties may be accommodated at all times." It is noted he had businesses in New York City until 1807, but it is not clear if he managed them from a distance in North Carolina or still lived in the city.

41. N. J. P., "Pilot Mountain".
42. "Tuttle's Hotel," *The Evening Post*, New York, NY, May 10, 1804.
43. "To Let," *The Evening Post*, Thursday August 22, 1805.
44. New York Council Minutes, 625.

THE MATHIEU FAMILY MOVES SOUTH

At some point after the summer of 1804, Andre Mathieu and his daughter, maybe his son, and new wife came down the Great Philadelphia Wagon Road with all their worldly possessions. This might have involved more than one wagon, although some of what he owned might have gone by ship to Charleston, SC. From there, the cargo could go over land to Salisbury, NC. It should also be noted that Mathieu owned land in Charleston in 1818 on Church Street going 26½ feet west and back—being 150-feet deep next to the Carolina Hotel.[45] When Mathieu traveled, he had a solid silver "hammered by hand fork and spoon" with him at all times.[46] Evidently, he had a fear of eating off pewter that might have had a very high lead content, which would leach into the human body easily. This especially happened to sailors of the French Navy, who then died from lead poisoning.[47] That was a good reason to avoid pewter in those days.

While a marriage certificate was not found in the colonial records, Andre Mathieu did marry a second time, and her name was Miss Frances Radcliffe.[48] It is said Miss Radcliffe was a rather temperamental artist and musician who returned to New York City from North Carolina. The second Mrs. Mathieu remained in New York City until her death on November 14, 1845. The funeral of Frances Mathieu took place at her son Andre's home at 97 Duane Street in New York City.[49]

The continuance of Mathieu's story is based mostly on newspaper articles, letters to the editors, deeds, and business records in New York City. His story picks up again in Salisbury, NC, where he became a Salisbury Commissioner along with John Giles on February 24, 1829.[50] He was then sixty-eight years old. There were no paintings, photographs, or journals of Mathieu to be found but maybe they are stored in someone's attic. There was also a "Letter to The Editor" in the Winston-Salem Journal by Lelia Graham Marsh, found at the Forsyth Library North Carolina Room. The letter was from a relative of Mathieu who corrected some misinformation in a story they ran but also confirmed that Andre Mathieu was mayor of Salisbury.[51]

There is a translation of a letter received in 1819 in Charleston, SC, from a cousin in France thanking Mathieu for some cotton that he shipped to them.[52] At the time Mathieu was a cotton buyer and when you said "shipped" back then it more often than not meant it literally went by ship.

---

45. Mathieu folder, 1966.
46. Mathieu folder, 1966.
47. Mathieu folder, 1966; "Early Pewter Was Beautiful, but Dangerous," George Way/Antiques, March 18, 2010, https://www.silive.com/homegarden/antiques/2010/03/early_pewter_was_beautiful_but.html.
48. Mathieu folder, 1966.
49. Mathieu folder, 1966.
50. News Item, *The Journal*, February 24, 1829.
51. Marsh, "Letter to the Editor," 1966.
52. Mathieu folder, 1966.

FIGURE 3.8. Is this an ancient dolmen from prehistoric times near the walkway to the Little Pinnacle? Are those symbols or just mold growing on the back wall? A dolmen is two parallel stones with a stone over both, which is commonly seen in Ireland and Scotland. We will delve into the various stone structures at Pilot Mountain State Park and their meaning in my next book.

In Salisbury, it was noted that Mathieu bought the northwest corner of the courthouse square, which was near a trade road and prime real estate at the time.[53] The store he had there was seventy-two feet in length, had five rooms, and a cellar. It featured a storeroom, counting room, and warehouse that years later, in 1842, sold for $4,500. His store was also situated near the Masonic Lodge on the east corner of the courthouse square. One time, "walking back from the post office with the letter in his hand," in Salisbury, he commented that his "mudder" had died and left him $10,000.[54]

Mathieu was involved with early Salisbury, which was the county seat for Rowan County. At one time, Pilot Mountain was in Rowan County, which obviously covered a vast area of land. There is an Andrew Mathew listed in the 1820 census for Rowan County with a female and both are listed over forty-eight years old.[55] On February 24, 1829, it was noted in The Journal, a newspaper in Salisbury, that Andrew Mathew was a Salisbury Commissioner.[56] Later, a newspaper

53. "Valuable Store House: For Rent," *The Fayetteville Weekly Observer*, March 4, 1830.
54. Obituary, *The Times Picayune*, December 23, 1857.
55. *1820 Census of North Carolina* (National Archives), 281.
56. News item, "Commissioners of Salisbury," *The Journal*, February 24, 1829.

account mentioned that Mathieu was an honored part of the volunteer Salisbury Fire Brigade.[57]

Mathieu's daughter from his first marriage, Teresa Hannah Mathieu, met and married a judge from Newberry County in South Carolina who went by the name of William Gillam (a.k.a., William Gilliam).[58] Teresa entered Salem Academy and College[59] when she was fourteen, but that institution might have gone by another name at that time. Teresa was teaching music and art in Salisbury at the time she met William Gillam.[60] Together they built the Gillam Hotel at Pilot Mountain in 1830, where they spent their summers, and then they wintered at their second home in Columbia, SC. The Gillams moved to North Carolina permanently in the following decade. Mathieu had wanted an entertainment house or inn at the base of Pilot Mountain since 1815.

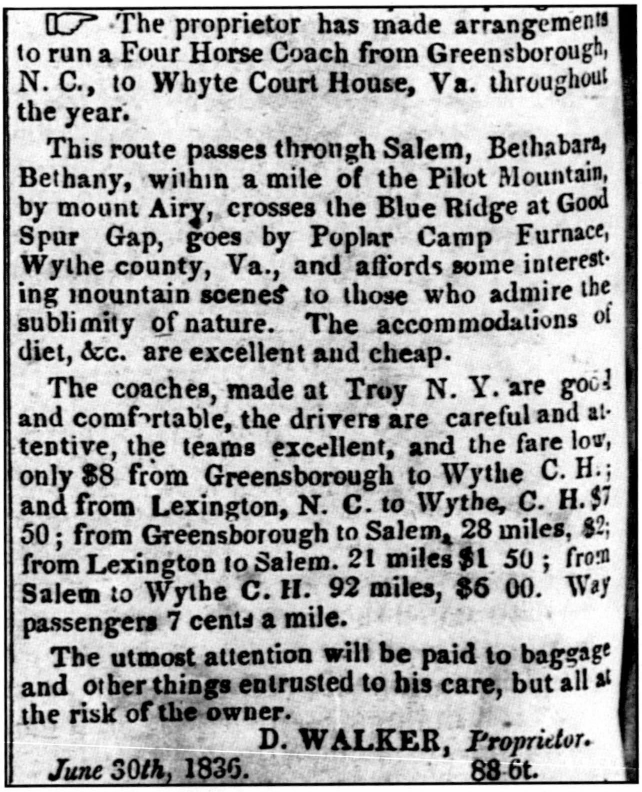

☞ The proprietor has made arrangements to run a Four Horse Coach from Greensborough, N. C., to Whyte Court House, Va. throughout the year.

This route passes through Salem, Bethabara, Bethany, within a mile of the Pilot Mountain, by mount Airy, crosses the Blue Ridge at Good Spur Gap, goes by Poplar Camp Furnace, Wythe county, Va., and affords some interesting mountain scenes to those who admire the sublimity of nature. The accommodations of diet, &c. are excellent and cheap.

The coaches, made at Troy N. Y. are good and comfortable, the drivers are careful and attentive, the teams excellent, and the fare low, only $8 from Greensborough to Wythe C. H.; and from Lexington, N. C. to Wythe, C. H. $7 50; from Greensborough to Salem, 28 miles, $2; from Lexington to Salem. 21 miles $1 50; from Salem to Wythe C. H. 92 miles, $6 00. Way passengers 7 cents a mile.

The utmost attention will be paid to baggage and other things entrusted to his care, but all at the risk of the owner.

D. WALKER, *Proprietor.*

June 30th, 1836. 88 6t.

FIGURE 3.9. An ad in 1836 for stagecoach rates going right by Pilot Mountain in The Weekly Standard.

57. "Mr. Editor," *Carolina Watchman*, March 6, 1851.
58. Mathieu folder, 1966.
59. "Our History," Salem College, https://www.salem.edu/about/history.
60. Mathieu folder, 1966.

The proprietor of the mountain, Mr. Mathieu, is, by the way, an old gentleman of more than four-score and ten years of age. He belonged to the French fleet under Count Rochambeau, which was sent to our assistance in the revolutionary war. Possessing his faculties unimpaired, he entertained us with an interesting history of his eventful life. He remarked that he dined with Gen. Washington in Boston shortly after the war, and saw Bonaparte at Toulon whilst he was yet an inferior officer in the French Army. When he started from the mountain he became our guide, and accompanied us on horseback to within a short distance of the pinnacle. Even at the advanced age of ninety-three he retains that vivacity and politeness which are characteristics of the French people. He reads the smallest types without the aid of glasses – May he will live yet many years to behold his favorite mountain and view with pleasure his noble forest, of which he is justly proud. At the base of the mountain is a strongly impregnated chalybeate spring. The accommodations for visitors do not compare very favorably with the elegance and sumptuousness of some of our northern watering places, but mine host, who is the son-in-law of the proprietor, does all in his power to render the stay of visitors agreeable and interesting. A large addition is being built to the Hotel, which will add to conveniences of the establishment.

—Staff Writer (1854)[61]

Given the exciting life that Andre Mathieu had before moving to North Carolina, it must have been fascinating to sit with him on a rock ledge on the mountain, riding horseback, at a dinner table, or on a porch to hear his stories. To be sure, his public life in New York City served him well for opening businesses in North and South Carolina. His mentorship to his daughter, who ran the Gillam Hotel for decades, clearly proved invaluable as this backcountry hotel was a success by all the newspaper accounts. Teresa probably kept a watchful eye on the silver after her father's experience in New York City at the Mathieu Hotel with the silver thief. Her husband William, after whom the hotel was named, was also a force in the process with a strong publicity campaign in the 1840s and 1850s.

I have been surprised that more persons and especially invalids, do not visit this place where a most excellent Summer Retreat is established by Wm. Gillam, Esq., who is a whole souled South Carolina gentlemen.

I have visited many of our public watering places am decidedly of the opinion that air and exercise have altogether more to do in restoring

61. Staff writer, "Original Sketches: Notes Along the Way," *Carlisle Weekly Herald*, Wednesday May 3, 1854.

> health of those who visit such places, then have the best mineral waters of any of our best public springs. In respect to exercise, balmy air, a fine opportunity to divert the mind and give a genuine zest to life, there certainly is no place superior, if equal to Gillamsville, the residence of 'Squire Gillam,' directly at the foot of Mount Ararat [Pilot Mountain] on the South side.
>
> —Published Private Letter (1849)[62]

One wonders why Gillam would decide to give up his city life as a judge in South Carolina, where his family had deep roots, for the country life in Pilot Mountain. There are no known reports of scandals, but one might conjecture that a judge did not always garner favor with his cases, and it might be that something tipped the scales in favor of a quieter life away from judicial duties in the city. In the next chapter, more will be shared on this time period from newspaper accounts about visits to Pilot Mountain in the early 1800s.

> The top of the pinnacle is nearly flat, and covered with stunted trees and under-growth, among the rest, blackberries and whortleberries – not equal, however, to the 'Big Blues,' for which Sampson is famous. Some enthusiastic person once preached on the pinnacle to a small but select audience. The Saura Town Mountains, some fifteen or twenty miles off at their nearest point, look within a stone's throw. It is said by some, that one of these, Cook's Peak, is higher than the Pilot, wherest old Mr. Mathieu – Mr. Gillam's father-in-law – is much displeased. The old gentleman is ninety-one years of age, yet active, lively, and conversable, with a marked French accent – although he came to the country about the time of the Revolution.
>
> —News Item (1853)[63]

The following was a guided tour of the mountain by Mathieu's three grandchildren, Dr. Gillam and his two sisters. This article contained the first mention of the card game to win Pilot Mountain, but it is unclear who in the family told the tale that was mentioned earlier in this chapter. He might have heard it from Mathieu himself, his daughter or son-in-law at the Gillam Hotel, or from his grandchildren on this trip climbing Pilot Mountain:

62. "At the Foot of Mount Ararat, Alias the Pilot Mountain," *Eastern Carolina Republican*, September 19, 1849, reprinted from *Danville Register*, August 17, 1849.
63. News Item, "Our Trip West," *The Daily Journal*, July 21, 1853.

THE PILOT MOUNTAIN,

A SUMMER RETREAT.

THIS GRAND CURIOSITY is situated in Surry county, two miles west of the Stokes and Surry line, and near the main road leading from Germanton and Salem to Rockford, 18 miles from Germanton and 25 miles from Rockford.

The Pilot Mountain House, 1½ miles from the pinnacle, has been newly fitted up for the approaching season, and no expense or attention will be spared to render enjoyment to visiters who may be pleased to seek health or pleasure at the Pilot Mountain.

One mile from the house is a fine mineral spring, and a good carriage road leading to it.

The ascent to the pinnacle has been rendered perfectly safe, and comparatively easy, by ladders made for the purpose. The view from this point is the grandest spectacle in North Carolina if not in the whole southern country, being a huge shaft run up 1800 feet into the blue etherial arch, from a comparatively level surrounding surface, and embracing the Blue Ridge and Allegheny Mountains, from the peaks of Otto to the north 80 miles, to the Table Rock in South Carolina, 100 miles including within the range Paris Mountain, Spartanburg district, and both peaks of King's Mountain in York district.

The large number of visiters to this grand curiosity has greatly increased within a few years, and the subscriber can assure the public that all who come shall enjoy themselves so far as depends upon him. The number of visiters during the last season, as shown by the register, was between three and four hundred.

A good Guide to conduct visiters up the pinnacle, and to point out the different and distant objects, will always be in attendance.

Travellers passing over the Blue Ridge, by Mount Airy, will lose but little in time or distance, by calling at Pilot Mountain House.

The purity of the air, the excellence of the water, and the beautiful scenery, never fails to invigorate the enervated and the invalid, and to make robust the man of health.

The plank road from Fayetteville west (now completed to Carthage) will go to Salem, within 25 miles of the mountain, and if extended, will probably go within a mile of it. Sixty miles of it will be completed this summer, bringing the traveller nearly half way to the mountain.

His prices are very moderate, as he wishes to encourage and induce the people of the State to visit this one of their grandest curiosities. WM. GILLAM.

Pilot Mountain, April 15, 1851. 865—tf.

PILOT MOUNTAIN.—Pilot Mountain I pass *ex route*. This mountain sometimes goes by the name of "The Ararat." Why called Pilot, no one seems to know, more than that it is supposed to have served as a Pilot to the Indians. But Pilot is not an Indian word.

It presents two separate views, wholly unlike each other. From the east it is conical, rising to a vast-height and terminating in a regularly shaped column, almost as perpendicular as masonry. This column goes by the name of "The Pinnacle." It is of rock and rises from the top of the mountain some three hundred feet. The top is a level surface of about three acres of earth and stones, with a good crop of half sized oaks and other trees upon it. A rabbit turned loose upon this level plateau would meet with difficulty in getting down again, owing to the steepness of the sides of the column.

There are several summer boarding houses in the immediate vicinity of the Pilot, and the wonder is that our country does not afford a sufficient number of the lovers of nature to keep these hotels filled with visitors during the summer.

The base of the east side is washed by the Yadkin river.

The South view is poetically called "The hog's back view," and appropriately, so far as outline goes. It has very much of a hog's back look about it. The Pinnacle, from the South side, is on what would be the hog's head.

Ascent to the top is not very difficult.—A carriage may be driven up the slope, say half the distance. Then the journey can be made a little further on horse back, but here the animals must be tied, and the "*Walker* Line" takes you to the base of the column. This you scale by a ladder.—But the summit once gained, you are amply paid for all your pains. The view is incredible on a clear day; I will not attempt a description of it.

Corres. Raleigh *Sentinel*.

FIGURE 3.10 and 3.11. An ad placed in the The *Weekly Standard* on Wednesday, May 7, 1851 (left). A feature article in the *Wilmington Journal* on October 18, 1867 (right).

This property was entered as vacant land by a company some fifty years ago, and afterwards sold to a mercantile house in the city of N. York, of which Mrs. Gilliam's father, a Frenchman by birth, was one of the members and to whom it was allotted in the division of their estate. Much of the land is first rate for tobacco, corn, and wheat, and but for its remoteness from market would be very valuable. On the North side, the mountain is covered with snow nearly the entire winter, and until the spring is far advanced.

On the next day after dinner, having become fully rested from the fatigues of my journey, in company with Mr. Dermill of Washington, N.C., (Traveling agent of the Mutual Insurance Company of Raleigh) young Dr. Gilliam and his two sisters set off for the mountain. We rode in our buggies a little more than a mile up a gradual ascent when it became so steep that we had to leave them, and commence the ascent on foot. We soon came to an excellent spring of water where we rested a short time and proceeded on our route. The path is winding and at places very steep and rocky, destruction to shoes of which mine soon gave evidence. At times we would pass some little land rich with growth of walnut and poplar. After much fatigue and frequent halts for relief principally of the young ladies, we reached the summit between the two pinnacles, the higher one on our right and the lower one on our left. After pursuing a winding path arrived to the North side of the highest pinnacle composed of a perpendicular mountain of rock. We commenced its ascent partly on ladders and then climbing between the chasms in the rock, formed by nature as if to invite the ascent of man. About half way up, a spring of water gently dropping down issued from the bosom of the pinnacle. Soon we were at its summit, with the cool zephyr of the evening refreshing and fanning our wearied bodies and imparting as it were new live and vivacity to the system. Our eyes were delighted with a little world beneath us over which we could view the distant mountains for a hundred miles and surrounding country with its farms looking like gardens, and its farm houses, villages, and the Yadkin with it rolling waves making its way through the wide valley with the luxuriant fields of corn and tobacco growing on either side. Far South was seen the famous King's mountain of Revolutionary memory as the scene of one of its severest battles. The view was sublime and delightful, and I feel that my feeble pen is wholly inadequate to its description.

On the top of the pinnacle there is an acre or more of table land quite rich, and there was on it a few years since a garden planted by the proprietor. On the rocks near the edges and on the trees many names of persons from different parts of the Union were engraved or cut. Among them I noticed a Miss Sarah Makepeace, all the way from Lynn, Mass. There were also names of many who have doubtless long since gone the way of the world and have left here, perhaps, the only memento of their remembrance. About twelve months since our esteemed brother Robertson, preached a sermon from a natural pulpit of rock between the pinnacles to near two hundred hearers. Near the West pinnacle there is a perpendicular wall which has the appearance of being built by hands, but is no doubt the handy work of the great Architect of Nature. When we had seen the sun set in the western horizon, long after he had become invisible to those below, we commenced our descent, which if anything was more tiresome than the ascent, and did not reach our comfortable quarters at Mr. Gilliam's until it was quite dark. As you may suppose we enjoyed a good supper, and after a short time retired to the shades of rest where nature's restorer balmy sleep, made us forget the fatigues of the evening scene, only when a wandering dream would carry us back to the mountain summit.

It is contemplated to establish a Female Boar[d]ing School at Mr. Gilliam's, which will be very desirable location for health beauty of scenery, and cheapness of living. I witnessed a painting by Miss Gilliam of the mountain which was very exact and beautiful. My sheet is full, and I must defer my account of the Iron Works and the County of Surry to a subsequent number.

—N. J. P. (1847)[64]

FIGURE 3.12. The view of the Gillam Hotel from the Little Pinnacle. The hotel was owned by Andre Mathieu's daughter, Teresa Hannah Mathieu, for nearly five decades.

"DEATH OF A HISTORICAL CHARACTER"

Andre Mathieu lived on Pilot Mountain in his later years and led tours as a guide around the mountain until he was ninety-two! He lived a long, exceptionally distinguished, full, and exciting life. For the last half of his life when he moved to North Carolina, he made it to the top of Pilot Mountain each year even in his

64. N. J. P., "Pilot Mountain."

ninth decade of life. This was a time period where a horse could only get you halfway up the mountain and the rest of the way was by foot. In 1855, the Mathieu corner in Salisbury burned shortly before Mathieu died. Fire was a huge threat to every city in North Carolina, and nearly all of the large cities burned at one time or another due to all of them being mainly built of timber. As his health declined, he moved to Davie County in the spring of 1857. Mathieu stayed with his granddaughter, Elizabeth Gillam Sheek, in Mocksville, NC, that summer and fall. Mathieu passed away at her home on November 9, 1857, at three o'clock in the afternoon.[65] In his will, Mathieu stated upon his passing: "To his daughter, T. H. Gillam for her separate use and not subject to control and debt of her husband, William Gillam, all stocks of horses, cattle, sheep, furniture, household possession of T. H. Gillam and her husband William Gillam at Pilot Mountain."[66] While the Gillams might have built the hotel, it appears from his will that it was furnished by Mathieu.

Mathieu, like President George Washington, was a member of the Masonic Freemason secret society and his funeral was officiated by the Mocksville Masonic Lodge at the grave site. For this book, the Mocksville Masonic Lodge graciously tracked down the handwritten log of Mathieu's funeral, which was also published in the newspaper, in their records from 1857. The Greensboro Patriot newspaper prefaced the Mason Lodge announcement of his death with the following:

> From the following proceedings of the Mocksville Lodge, it will be seen that another Revolutionary Soldier is gone – Thus they continue to pass away, until the number of those who in 76, endured so many trails and hardships for the blessings which we now enjoy, is quite small. In a few years more, the last survivor of that noble land, will have gone to receive his reward. Mr. Mathieu, was a Frenchman, he was a highly respectable gentleman, a good citizen, and his memory will long be cherished.[67]

The Mocksville Mason Lodge then made note of his passing in their hand written records as follows the week of his death (that was also published in The Greensboro Patriot):

---

65. "Deaths," *The Spirit of the Age*, November 25, 1857.
66. "Andre Mathieu Will of June 11, 1855," in *Davie County Will Book 1*, Davie County Public Library, Mocksville, NC, 158.
67. Handwritten meeting notes regarding the passing of Andre Mathieu, Mocksville Freemason Lodge #134, November 11, 1857; "From the Following Proceedings of the Mocksville Lodge," *The Greensboro Patriot*, November 27, 1857.

> Lodge opened in Masters degree:
>
> The object of the meeting being to make the necessary arrangements to attend the burial of Bro Andrew Mathieu today at 12 N at Olive Branch
>
> The Lodge was called from labors to refreshment seven o'clock P.M. Lodge resumed labors when the following resolutions were offered and unanimously passed.
>
> It has pleased an all-wise providence to remove from our midst on the 9th our aged and esteemed Bro Andrew Mathieu who was a Revolutionary Soldier aged about 96 years.
>
> Res'd, that in his death masonry has sustained the life of a worthy Bro and citizen. Res'd that we recognize in this painful bereavement the uncertain and precocious tenure of our existence, and admonition to live up to the tenets of our order. Re'sd, That we tender to the family of the deceased our heartfelt sympathy.
>
> Res'd, that a copy of these proceedings be forwarded to his daughter, Mrs. Gilliam, and to the Salisbury Watchman for publication.
>
> No further business coming before the Lodge it was closed in due and ancient form.[68]

"Death of an Historical Character" was the headline of Mathieu's (a.k.a., Andrew Mathew) obituary, one of four found of five published, in a New Orleans paper on December 23, 1857.[69] A few years before his death, on October 10, 1850, Mathieu formally, "deeded a large amount of land in Surry County, and other land in Salisbury, in Columbia, SC and in Charleston, SC to his grandchildren: Lewis M Gillam of Surry Co.; to Juliet L Bernard; to Marietta, William and Cornelia Gillam," according to Lelia Graham Marsh in the Winston-Salem Journal.[70] This information was all supplied to Marsh by a great-great-grandson of Mathieu named John L. Davidson of Quitman, GA.

68. "Deaths," *Spirit of the Age.*

69. "Obituaries," *The Greensboro Patriot,* November 27, 1857; "Obituaries," *The Pittsburgh Gazette,* December 18, 1857; "Obituaries," *The Times Picayune;* "Deaths," *Spirit of the Age.*

70. Marsh, "Letter to the Editor."

FIGURE 4.1. Stone face on a Pilot Mountain State Park trail.

4

Early Documentation, Observations, and Access in 1815 and 1823

FIGURE 4.2. See the huge stone face winking on the far left of Pilot Mountain State Park on the south side of the mountain only seen from a great distance? One wonders where the stone circle would be for the smoke signals on the Big Pinnacle seen above. This is a teaser here for the second book of the series, which goes into more detail about the stone faces seen in the first page of each chapter.

As we learned in chapter 2 on the Moravian journals, the area around Pilot Mountain was self-sufficient other than salt. They used to import salt by the wagon full. While the area was isolated because of the lack of transportation by any means, be it water, railroad, or good roads, people got by. Not only did they thrive but diverse groups got along with each other.

"The period from 1790 until the 1820s were the halcyon days for Surry," according to The Heritage of Surry County, North Carolina, noting those happy days from the past.

> It was the land of plenty in the context of the times and even in the context of the 1980s. Wood for building and for fuel was limitless. Nature provided abundant food. Wild game including quail and wild turkeys were there for the taking. According to information found on tax lists, every family had cows. And cows provided milk, butter, cream and cheese. There was also something called clabber; this was a milk solid, soured and skimmed of cream. It could be eaten with a spoon and was very much like yogurt. Cows also proved fresh and dried beef and hides for leather.
>
> Every family grew sheep for wool and for food. Flax and some cotton were grown. Linen thread from the flax as woven with wool thread to make a material called linsey-woolsey.
>
> —The Heritage of Surry County, NC (1997)[1]

FIRST NEWSPAPER ACCOUNT VISITING PILOT MOUNTAIN IN 1815 FOUND

On August 19, 1815, a Dr. Jeremiah Battle of Edgecombe, sent a letter to General Calvin Jones of Raleigh giving an account of visiting Pilot Mountain on a long trip across the state. This description is the earliest long-feature published account found of being right on the mountain for the general newspaper reader. The Moravians had written about the mountain for nearly six decades in their journals before this account was published. This letter became iconic, often being quoted and reprinted in newspapers. The last full reprinting of the long letter was found in the Saturday edition of the *Tarboro Press* twenty-nine years later on June 29, 1844. Likely so much more was seen and discussed on Pilot Mountain than was eventually published in these accounts. Unlike in modern times, for the time period described in this chapter, there was no intentionally restricted access to these very interesting accounts by highly educated men. The purposely veiled

1. *The Heritage of Surry County, North Carolina*, Vol. I (Alleghany County Coordinator, 1997–2010), as quoted by NCGenWeb Project, https://sites.rootsweb.com/~ncsurry/history.html.

aspects of Pilot Mountain's history by law will be addressed in detail with documentation in the second book of this work.

Dr. Battle stayed overnight about four miles away from Pilot Mountain but mentioned seeing the mountain on the journey toward it from about seventeen miles away. "This curiosity of nature," Battle began, as he turned his attention in the middle of his letter to Pilot Mountain on his trip. Locals today have seen this blue appearance the mountain takes on during the day all their lives.

> We took the road which led by Mr. Alrord's, where we stopped and dined; here we had a most sublime and interesting view of the mountain, which exhibited a blue appearance, rearing its head of rock many hundred feet above the tops of the adjacent hills and trees. Its shape at this distance had a striking resemblance to that of your market house, the roof of which representing the base and body of the mountain; the perpendicular octagon, the main pinnacle of rock, and the shingled top of this, the elevation of earth covered with rude heaps of rock, shrubs and trees on the top of the rock pinnacle.
>
> —Dr. Jeremiah Battle (1815)[2]

Following a dinner, they were entertained by their hosts, Mr. Shepard and his lady. Shepard lived off the road about four miles from the mountain and one wonders if he was related to one of the owners by that name (spelled slightly differently) in the 1700s. The very next morning they engaged a guide, John Fletcher, to show them the way to Pilot Mountain and "conduct [them] thither." Just like now, as they slowly approached the mountain, its color changed from a strong blue hue to dark green. As many had done before and after in the 1800s, they made it with ease to the mineral spring on the south side by horseback. At that point you would leave your horse at the mineral spring before it became too steep to continue on horseback. From there climbing was difficult, unless you used the switchback method of rock climbing that the guide would show you. It was then your choice to follow your guide's instruction or do an exhausting and humbling crawl straight up the mountain on your hands and knees. Much later in the nineteenth century, there was a rope to hold onto for those walking up the mountain on the south side from the mineral spring onward called "the walkers line."

> When we were yet a mile off, we stopped and viewed it again with increased delight; the former blue appearance changing into a rich dark green, of decorating trees and shrubberies, thro' which we could perceive its rocky surface; the pinnacle of granite rock, the white appearance,

2. Jeremiah Battle, "The Pilot Mountain: Extract of a Letter from Dr. Jeremiah Battle, of Edgecomb, to Gen. Calvin Jones, of Raleigh," *The North Carolina Star*, September 29, 1815.

seemed a magnificent building, ornamented with green briars, vines and shrubs, at suitable distances, growing out of its wall, combining in an eminent degree the sublime and beautiful.

Whilst we were ascending on foot, and experiencing the fatigue of rising step by step, on an angle of about 45 degrees the weather being hot, I recollected that my neighbor Mr. W. some years ago, having labored under chronic disease and debility, and excited by desires like my own, lost his life by an acute inflammation, induced by precisely the expedition I was now performing. I took the precaution of going in a zigzag direction, which rendered the ascent much easier. We then turned our course, took an obscure pathway which led along a ridge probably 100 feet above the vallies on each side the hard effort was immediately worth it once the ascent was completed and the view revealed to the climber.

—Dr. Jeremiah Battle (1815)[3]

FIGURE 4.3. The hue of Pilot Mountain State Park seen from downtown Mt. Airy.

Just like what happened to so many who traveled up the mountain, a rainstorm was seen approaching in the distance toward the group that Battle was with. They were "compelled" to seek shelter under one of the many rocky open

3. Battle, "Pilot Mountain."

FIGURE 4.4. An image taken from an airplane above the Big Pinnacle with the trails around it clearly seen. The top does seem like a flat layer cake that gets smaller as you go to the top. There seems to be a trail that goes right across the middle of the Big Pinnacle, which might have led to a stone circle for the smoke signals and fire beacons.

caverns to stay dry. These same prehistoric overhangs are mentioned in the General Management Plan of 2018 in a theme table by Pilot Mountain State Park. That controversial prehistoric theme table will be addressed in detail in the next book. After that brief shower, they had to decide if they were going to summon up the courage to go up to the top of the Big Pinnacle or not.

> Our pilot ascended the pinnacle, but we had not the enterprise to follow him up this steep wall of rock 300 feet high. It is perpendicular or projecting over on every side for 275 feet in height, except one narrow steep passway, with slight footsteps; a one part of which, about 30 or 40 feet high, the rock is perpendicular for 4 or 4 ½ feet high from one step to the next-here it takes an expert climber to make his way without assistance; afterwards the ascent is easier. . . After the rain had ceased, and I had become more familiarized with the activity at first so awful, I resolved to follow my guide and fear no evil; which I did with little difficulty, except at the place mentioned above.
>
> —Dr. Jeremiah Battle (1815)[4]

4. Battle, "Pilot Mountain."

Were these footholds described by Battle built into the Big Pinnacle by the Native Americans to give easy access to send smoke signals and observe the cosmic calendar? Or were these steps designed by one of the four earlier civilizations found in the oral history of this land? In other words, once again, the oft-repeated theme returns: are they natural or manmade? Battle reported that his guide noticed that more women than men had ascended to the top of the Big Pinnacle. He also reported that one man who had come from a considerable distance was forced to the top by his "cruel comrades" who "endangered the faculties of his mind." Battle was also impressed with the local people around Pilot Mountain and concluded his published letter with the following:

> On the slightest interview with the inhabitants, I found in the enjoyment of blessings which no money can purchase, viz; health and contentment. The country is peculiarly calculated to form strong and robust constitutions, with bold enterprising dispositions. We need not wonder, therefore, at the praises acquired by our mountaineer soldiers, in the great and desperate enterprises in which they have been engaged.
>
> You would suppose from the account I have just given you of this rude display of nature, that it is wholly destitute of the means of benefitting either man or animal. But it affords good grazing. The grass springs up between the two rocks in such abundance that it was contemplated by the French gentleman, on whom it had been imposed as a tract of good land, to form this mountain into a sheep walk, and he actually placed on it 100 head of sheep; but he had a bad shepherd, who suffered them to die for want of winter feeding, which disgusted the owner, and he gave up his plan. He also intended to establish a house of entertainment, at the above-mentioned Mineral Spring, and invite valetudinarians to resort thither for the recovery of health. The Spring being at the foot of Pilot Mountain, would tend greatly to attract visitors. With much esteem, I am Sir. Yours.
>
> —Dr. Jeremiah Battle (1815)[5]

1823 SCIENTIFIC STUDY OF PILOT MOUNTAIN IN NEWSPAPERS

In 1823, two professors and the first president of the University of North Carolina at Chapel Hill (UNC-CH; which opened its doors in 1795) traveled to Pilot Mountain to do a scientific study of what Battle had earlier in 1815 called, "a

5. Battle, "Pilot Mountain."

FIGURE 4.5. Have the trails changed that much from historic times in the 1800s or even the prehistoric times?

natural curiosity."[6] Their names were Professor Denison Olmstead, Professor Ethan Allen Andrews, and President Joseph Caldwell.

Professor Denison Olmstead (1791–1859) was a physical scientist and professor of chemistry starting in November of 1817 at UNC-CH. Olmstead was credited with doing significant pioneering work in the geology of North Carolina. Olmstead was hired by the Board of Agriculture to do a four-year mineralogical and geological study of North Carolina. For that he was paid $250 per year. When his report was published for the Board of Agriculture it was believed to be, "the first probably of any State in the Union."[7] Olmstead was the one who wrote the article but it wasn't published until 1826. There might have been an earlier version of the Pilot Mountain account, but it wasn't found.[8]

> What must have been the feelings of the first scientific party to scale the slopes of Pilot Mountain, an eminence which has intrigued the interest of both scientist and average citizen from the days of earliest settlement? Although relatively unimportant with respect to its elevation (when compared to great mountains elsewhere in the world) the mountain has long been looked upon as one of the really great wonders of Eastern United States.

6. Battle, "Pilot Mountain."
7. Kemp P. Battle (Kemp Plummer), *History of the University of North Carolina*, vol. 1, *From its Beginning to the Death of President Swain, 1789–1868* (Raleigh: Edwards & Broughton Print. Co., 1907), 289, https://docsouth.unc.edu/nc/battle1/battle1.html.
8. "Documenting the American South," 2005, UNC-CH University Library.

One of the most early descriptions of the Pilot and one which possibly was the first to be written by a trained geologist and educator was based upon a scientific excursion made to the top of the mountain in the summer of 1823 by celebrated Professor Denison Olmstead, who was chosen professor of chemistry, minerology and geology at the University of North Carolina [Chapel Hill] in 1817.

Professor Olmstead, who was an American physicist and astronomer, was born at East Hartford, Conn., in 1791. He was graduated at Yale University in 1813, studying theology for a time and finally turning to science. He came to the University of North Carolina in 1817 and after thoroughly successful work there, accepted the chair of mathematics and natural philosophy at Yale from 1825 to 1836.

Olmstead and Elais Loomis were the first scientists in America to make observations on Halley's Comet upon its return in 1835 and in addition his views on this great celestial wonder, he published numerous scientific works.

More than ordinary interest, then, attaches to the fact that he and President Joseph Caldwell, of the University of North Carolina, together with Professor Andrews, also of the University faculty, surveyed the mountain and recorded what they found for later generations, possibly the first such survey every made there.

—Harvey Dinkins (1923)[9]

President Joseph Caldwell (1773–1835) was a Presbyterian minister, mathematician, astronomer, and the first president of UNC-CH. It is interesting to note that Caldwell was on Pilot Mountain in 1823. Clearly, we know now that Pilot Mountain is chock-full of astroarchaeology bell ringers with its natural east-west lineup. This visit by the academic trio was two days after the summer solstice on June 23, 1823. This would be the day that the sun began to move again after its standstill on the horizon for the seventy-two-hour period of the solstice. Was the date they visited Pilot Mountain random chance or on purpose?

The third member of Olmstead and Caldwell's academic team, Professor Ethan Allen Andrews (d. 1858), taught Greek and Latin at UNC-CH from 1822 to 1828. Andrews had graduated with honors from Yale in 1809. In later life, he wrote textbooks in Latin for students. The year before the team's study, in 1822, Caldwell had published a book on geometry. Olmstead had also published papers on North Carolina geology in 1820 and 1822.[10]

9. Harvey Dinkins,"Great Geologist Made Study of Pilot Mountain in 1823: Professor Denison Oldmstead, Famous in the United States 100 Years Ago, Gave Natural Wonder in Surry County Careful Study," *Winston-Salem Journal*, 1923.

10. Battle, *History of the University*, 1:289.

FIGURES 4.6 and 4.7. Two examples of woodcut blocks of Pilot Mountain used throughout the 1800s in newspapers before photography was invented. The hogback outline from the south is clearly seen on the top and the almost perfect pyramid view from the east seen on the bottom.

Olmstead wrote the feature "The Pilot Mountain," but it isn't clear where it first appeared in a newspaper over the course of thirteen years of it being published since their visit.[11] One summary from 1844 combines Battle's 1815 visit

11. Denison Olmstead, "The Pilot Mountain," *Raleigh Register and NC Gazette*, November 3, 1826; Denison Olmstead, "The Pilot Mountain," *The Patriot*, November 15, 1826; Denison Olmstead, "The Pilot Mountain," *Carolina Observer*, July 2, 1829; Denison Olmstead, "The Pilot Mountain," *Weekly Chronicle* and *Farmer's Observer*, June 18, 1836.

FIGURE 4.8. This pyramid image facing west of Pilot Mountain State Park is a modern attempt, taken March 20, 2022, to recreate what Professor Denison Olmstead saw at sunset in 1823.

and Olmstead's 1823 visit. This summary featured an oft-used pyramid shape woodcut block of Pilot Mountain for artwork.[12] In 1914 someone referenced almost all of the 1823 newspaper article for The Charlotte Observer[13] and Harvey Dinkins did a summary in the Winston-Salem Journal and Sentinel around the same time period in 1923. [14]

The academic team commented on the symmetry of Pilot Mountain in 1823 and raved about its beauty as they watched a sunset highlighting the outline of the mountain. Did the professor know of the sacred nature of the mountain, or did he pick that up intuitively? Olmstead's description of the pyramid-mound aspect of Pilot Mountain was interesting. This was a keen, early observation not picked up by those coming from the south or the north of the mountain, as it looks different from each direction. This type of observation would not enter academic circles until three decades later as a feature of the Mound Culture by the Smithsonian Institution. Much of the academic team's report on Pilot Mountain is of interest and is quoted here, letting our ancestors speak, but off-record accounts would be just as interesting:

12. Staff writer, "The Pilot Mountain," *The Greensboro Patriot*, April 13, 1844.
13. Staff writer, "The Art and Literature of Earlier Times as Applied to One of the Natural Wonders of North Carolina—Historic Pilot Mountain and Its Peculiarities," *The Charlotte Observer*, February 1, 1914.
14. Harvey Dinkins, "Great Geologist Made Study of Pilot Mountain in 1823: Professor Denison Oldmstead, Famous in the United States 100 Years Ago, Gave Natural Wonder in Surry County Careful Study," *Winston-Salem Journal*, 1923.

FIGURE 4.9. A mineral spring drains down the south side of the mountain, which is the way Dr. Battle, Professor Olmstead, President Caldwell, and Professor Andrews went up the mountain. When people came to the source of the mineral spring, they would rest here since it is halfway up the mountain.

In the first glimpse we catch of the Pilot in Rockingham, it resembles a magnificent temple with a superb cupola, not unlike the picture of St. Peter's at Rome. The uncommon symmetry of its structure is preserved on a much nearer view. Nothing could exceed the regularity and beauty of its appearance, as it presented itself to President Caldwell, Professor Andrews, and myself, on a summer evening of 1823, as we were approaching it from the east, a little before sun-set. Its dark side being towards us we could more distinctly observe its finished outline, which was still illuminated. The figure now presented by its sloping sides and perpendicular summit, was that of a triangle, having a portion of its vertex removed and replaced by a parallelogram; while the trees and shrubbery that graced the outline, appeared like delicate fringe projected on the western sky.

—Professor Denison Olmstead (1829)[15]

15. Denison Olmstead, "The Pilot Mountain," *Fayetteville Weekly Observer*, July 2, 1829.

Clearly, this trio of academics came well prepared with instruments to measure and examine Pilot Mountain for themselves. They awoke the next morning near the base of the mountain, still on the eastern side. They again would have seen the pyramid shape to the mountain but this time in the blazing morning light shining on the Big Pinnacle like a spotlight among the huge trees. While it is difficult to get a viewpoint, there is a sunrise pyramid shadow just as there is a sunset pyramid shadow at Pilot Mountain State Park. It's possible that Olmstead saw the pyramid shadow on the group's early morning hike the following morning.

Like Dr. Jeremiah Battle and his group eight years before, the trio were guided up the mountain. This time, however, their guide up the mountain was their host from the evening before. At an early hour, they made their ascent up the mountain. As they got closer to the Big Pinnacle, the mountain reminded Olmstead of the Seuir of Eigg in the Western Island of Scotland with its "almost perfectly cylindrical" shape. As another comparison, the natural crystal capstone of the pyramid-shaped Pilot Mountain is similar to what used to be on top of the Great Pyramid in Egypt.

> Dr. Caldwell and Professor Andrews had provided themselves with a quadrant and a mountain barometer, for taking elevations, while I was to examine the Geology of the mountain. A small stream called Grassy Creek, which runs southerly, being considered as the true base, at this point we began our observations. For more than half the distance from this spot, the ascent is so gradual, that one may proceed on horseback, the acclivity being only about 20 degrees until we reach the Spring, a post of refreshment which was very grateful to our party.
>
> The water was very cool and pure, its temperature being only 58 degrees (June 23d) which may be regarded as the mean temperature of the place for the year. From this spot the ascent becomes more abrupt, (about 25 degrees) and those who are unaccustomed to climbing mountains find it extremely fatiguing. We arrived at the pinnacle on the north side, where is the only pass that has hitherto been found to the summit... The perpendicular wall is 200 feet in height; and many of the visitants, unaccustomed as they are to Alpine scenery, are so affected by the bewildering aspect of the world below them, and so appalled at the idea of hanging on the sides of the cliff that frowns over their heads, that no persuasion can induce them to ascend the pinnacle.
>
> —Professor Denison Olmstead (1829)[16]

Just as Battle had experienced, this team found it daunting to proceed up the last two hundred feet to the top of the Pilot Mountain. This was after the long

16. Olmstead, "The Pilot Mountain," *Raleigh Register* and *NC Gazette*.

climb to the saddle area from the mineral spring area. It was a time before ropes, ladders, stairs, or handrails had been installed to climb up the mountain. However, it was the only way up for humans and animals alike, so they had to face their fear and climb up using the handholds and footholds on the Big Pinnacle. Reportedly, coming down was the most terrifying part, even when there was a ladder or stairs. If you hang your head out of a fifteen or twenty-story building, you will get an idea what that is like to climb up or down. When you got to the top, you were faced with the prospect of Timber rattlesnakes and various types of hawks, but the view was, and is, incredible. Since access to the top of the Big Pinnacle is now denied by Pilot Mountain State Park, because of the rare plants and the nesting ravens, it makes this account all the more fascinating.

> The path is indeed narrow and steep; but it appears, when viewed from below, more formidable than it really is. In some places, the ascent is nearly perpendicular; but convenient cavities and projections are found, by which the feet and hands may be made sure. We were too much engrossed by the scenes that expanded around us, to proceed with our professional tasks, but seated ourselves on the northwestern brow of the pinnacle to enjoy the sublimity of the prospect. The air was still, but a hollow roar ascended from the plain – the voice of the forest – and not less sublime that the roar of the ocean, which it seemed to emulate.
>
> —Professor Denison Olmstead (1829)[17]

Also like Battle, the party was watching the weather changing, but these were the days when you just watched it approach. The weather was predicted based on the barometer, how you felt, and how the animals were reacting, but for the most part it just approached you."Red sky at night, sailors delight, red sky in the morning sailors take warning" was a familiar predictive phrase. Olmstead commented on the weather:

> As the day advanced, these clouds began to multiply on the sides of the Blue Ridge, covering it acclivities with chequered fields of sun and shade. A few of them occasionally wandered towards us over the clear blue sky, projecting their dark shadows on the earth, which coursed each other majestically over the sunny tops of the hills and forests. At length, here and there a cloud rose above the Blue Ridge, and distilled a copious shower of rain, as it moved along the mountain from west to east, the exact limits of which we could easily define, the sun still shining on all regions around. Each successive cloud diverged farther and farther to the east, until a shower, accompanied with lightning and thunder, was

17. Olmstead,"The Pilot Mountain," *Raleigh Register* and *NC Gazette*.

> approaching the Pilot, and forced us to descend from the pinnacle, and take shelter under one of it shelving rocks. Here we had leisure to exchange our expressions of delight and admiration; and some of the party who had viewed the scenery in populous and cultivated regions that was more beautiful, still acknowledged that they had never witnessed any that contained more of the elements of the true sublime. Serenity was shortly restored to the sky, and we proceeded with our respective tasks.
>
> —Professor Denison Olmstead (1829)[18]

This esteemed UNC-CH team was most likely the first to take a scientific look at Pilot Mountain, but there are other detailed Moravian accounts found in chapter 2. It should be noted that mica was the prized possession of the Mound Culture. North Carolina mica was second only to native copper of the Keweenaw Peninsula from Lake Superior. Native copper from Michigan is so pure it reset the standard for purity of copper for the world. Here is what Olmstead said about the geology of Pilot Mountain in 1823:

> While my companions were employed in these observations, I had begun an examination of the geological structure of the pinnacle. A footpath running close to its base, conducts one, without the least obstruction, quite round the circle, and no opportunity could be more favorable for remarking the different kinds of rocks and their relative position. In the geology of the pinnacle, there is something quite remarkable and curious; and the geologist will linger around its base with as much delight and admiration, as he gazes upon the landscape from its summit. The pinnacle is made up chiefly of mica slate and quartz; but each exhibit peculiar and interesting characters. Its rocky wall is full of rents from top to bottom, and it is also regularly stratified, the strata dipping easterly, at an angle of only ten degrees. By these parallel seams, the whole is divided into tabular masses
>
> —Professor Denison Olmstead (1829)[19]

Tabular masses can often be seen in the walls of road cuts. They are called tabular because they are shaped like slabs or tablets. The team's geologic observations stand today.

18. Olmstead, "The Pilot Mountain," *Raleigh Register* and *NC Gazette*.
19. Olmstead, "The Pilot Mountain," *Raleigh Register* and *NC Gazette*.

FIGURES 4.10 (above) and 4.11 (below). Courtesy of Frank Jones Collection Forsyth Public Library. Courtesy of the Mount Airy Museum of Regional History inside the train station display for the Cape Fear and Yadkin Valley Railroad.

The most abundant rock is a peculiar kind of mica or grit rock, composed of very fine granular quartz with flesh red mica intimately disseminated. The texture is exquisitely fine, and the cohesion is so loose that it may be frequently crumbled between the fingers into the finest white sand. At a mill near the river Ararat, I saw a pair of mill stones, said to have been quarried from an eminence on the northwest side of the Pilot. They

> consisted of quartz rock, somewhat resembling French Burrh, and appeared to be of excellent quality. Grind stones are also quarried from the grit rock of these mountains.
>
> —Professor Denison Olmstead (1829)[20]

These two different detailed newspaper features, one by a respected doctor and the other by the professors at the UNC-CH, brought great attention to Pilot Mountain throughout the state. That description of the mica of Pilot Mountain makes one wonder if it's what is found throughout the Mound Culture as part of burials a thousand miles away from the mountain. Of course, along with the mounds listed in North Carolina, there were many mica mines listed in various counties of the state. It is believed that all mica found in the Mound Culture was sourced from North Carolina. Why was mica so valued? Mica replaced looking in a hole in the ground into still water that was used as a mirror. Your true reflection could be seen in mica. Looking at the still water in a hole in the ground you would only see your silhouette unless seen during a moon bright evening. To add to the mystery, there is a trail on Pilot Mountain called the Grindstone Trail, which might have had a prehistoric use. There is an example of a grindstone from Pilot Mountain in the visitor center.

After these accounts were published, the roads slowly got better in the state. Salem became a prominent trading center in the late 1700s and early 1800s. The pioneering Moravian physician, who was the only one around for a hundred miles, finally had help. Also, their grist mill was heavily used from the day it went into operation, and the Great Philadelphia Wagon Road came right through the area. Those trails are noted on the Surry County map by Fred Hughes, which is in book two of this series.

> The courthouse was usually the center of each town with the stock and whipping post occupying a prominent place in the courthouse yard. The town as a rule had but one street worthy of the name. At one end, just opposite the courthouse, were the stores and shops, while spread out along the length of the street were the homes of the most prosperous citizens set a few yards back in groves of trees. Every self-respecting town of at least five hundred inhabitants contained a tavern, five or six retail stores, a blacksmith's shop, and perhaps a shoe shop, a church or two, and a male or female academy which was 'situated eligibly, and neatly appointed, upon lots purchased by the citizens.' The larger towns usually had a public market where country produce was brought for sale. When

20. Olmstead, "The Pilot Mountain," *Raleigh Register* and *NC Gazette*.

FIGURE 4.12. Another 1800s newspaper with the oft-used woodcut block depicting the Gillam Hotel and Pilot Mountain. The Gillam Hotel plot of land on deed maps was noted as being 207.7 acres on plot number nine.

this was not the case, the courthouse yard or the street in front of the courthouse served the purpose.

—Staff Writer (1928)[21]

MOVING TO MORE MODERN TIMES AROUND PILOT MOUNTAIN

The railroad access for travel was late to arrive in the backcountry of North Carolina, and once it did, there was some criticism due to schedules that brought tourists by the mountain at night. For a long time, North Carolina was known as the "Rip Van Winkle State." In 1830, it was accessed by a legislative committee as, "a state without foreign commerce, for want of seaports or a staple; without internal communication by rivers, roads, or canals; without a cash market for any article of agricultural product; without manufactures; in short, without any object to which native industry and active enterprise could be directed."[22] That all changed with the railroads that arrived in the Pilot Mountain area in 1888, slowly snaking across the state.

21. North Carolina Historical Commission, "The Ante-Bellum Town in North Carolina," *The North Carolina Historical Review* V, no. 1–4 (January–October 1928), 374–375.
22. Wiley J. Williams, "Rip Van Winkle State," NCPedia, 2006, https://www.ncpedia.org/rip-van-winkle-state.

> Prof. J. A. Holmes, State Geologist and likewise Professor of geology at the State University went up the C.F. & Y.V.R.R., to Pilot Mountain, last Friday, with a company of students advanced in this study, fully equipped for work, and Prof. Holmes will spend some time imparting instruction to the young gentlemen in this particular line.
>
> —Staff Writer (1890)[23]

FIGURES 4.13 and 4.14. The people here are wearing their Sunday best. These images were found in an attic area circa 1900 by a friend of Betty Gay Shore. There must be other images like this of people on the ladders of the Devil's Tower in Wyoming in the same time period.

23. Staff writer, "University Students Abroad," *News and Observer*, June 14, 1890.

Later, more was published on the students going to Pilot Mountain with Professor Holmes:

> The attempt to paint, in words, anything like a correct picture of the view had from the Pinnacle of the Pilot Mountain must prove a sad failure, even if the pen were wielded by the hand of a 'ready writer.' A good landscape painter might possibly succeed in transferring to canvass an imitation – which would faintly resemble the original – but to be fully appreciated it must be seen.

FIGURE 4.15. Another image, circa 1900, that displays the rickety ladders to the top of the Big Pinnacle. Courtesy of the Charles H. Stone Library, Pilot Mountain, North Carolina.

FIGURE 4.16. This image is courtesy of the Surry County Historical Society and was used on the cover of Carolyn Boyles' book, *Early Days of Pilot Mountain, North Carolina: A History and Genealogy*. This might be the most well-known image in the entire book series. The woman with the hat on is believed to be an early image of Nevada Coleman, who lived on the mountain for twenty-eight years and was the wife of the one of the many caretakers, Elbert Bertis (Bert) Coleman.

All the toil and fatigue of the ascent is for the moment forgotten as soon as you place your foot on top of that magnificent pile of Omnipotent masonry. It is so very different from what you have been accustomed to that makes you almost dizzy – and at first it presents itself to the mind as something unreal, visionary and unsubstantial. I know I felt like I should not have been at all astonished if the whole thing had vanished from my sight in a moment, and left me standing 'solitary and alone' in one of my native haunts in the plain. But you gradually recover from this sensation, and begin to survey, one by one, the novelties which surround you. Just as far as the eye can reach in every direction there is something new, beautiful and sublime to arrest attention. Away to the west, you can trace the range of the hazy Blue Ridge – and even there recollections and reminiscences of the past are called up by looking at, it – for you go back to the time when you traced that same range in a map, surrounded by innocent, infantile and happy faces. And you also go a short ways into the future and reflect that there that range will still stand when all those school-mates and you and I shall have gone to that 'undiscovered country from whose bourn no traveler returns.

—Pee Dee (1858)[24]

FIGURE 4.17. Thousands of people looked for this North Carolina landmark moving south in the 1700s. Image taken March 20, 2023, on the equinox from Virginia.

24. Pee Dee, *The Greensboro Times*, October 9, 1858.

FIGURE 5.1. Stone snake on a Pilot Mountain State Park trail coming out of a dolman?

5

Finally, the Road to the Top of Pilot Mountain in 1929

FIGURE 5.2. An image of W. L. Spoon clearing the timber for the road up Pilot Mountain in 1929. Courtesy of the Wilson Library at UNC-CH and the W. L. Spoon Collection.

After Andre Mathieu passed in 1857 around the age of ninety-seven (his exact birthdate was not found in research for this book), his daughter, Teresa Gillam, continued running the Gillam Hotel with her husband, Judge William Gillam. They had six children together. During the Civil War, William died on February 8, 1862. The years during the Civil War must have been particularly difficult with cannon fire heard in the distance. As a child, Carolyn Boyles, the Pilot Mountain town historian, listened to the firsthand Civil War accounts told sitting on the porch in the evening. She said that at the end of the war the soldiers just walked home and were often barefoot. Pilot Mountain inhabitants never knew who was in the barn in the morning, as so many sought shelter and looked for some eggs to eat.

> History: Exploitation of the resources and scenery of Pilot Mountain has been attempted by its several owners with but meager success. At times every tree of merchantable size has been cut for lumber. Recently, however, the only cutting has been of cordwood from around the base. Several orchards and cleared fields are found on the lower slopes, and there are indications that a number of others once existed but have been abandoned. As recently as 1930, a field was cleared on the top of the The Ledge, but it was abandoned after a few years of unsuccessful cultivation.
>
> Teresa Gilliam, for whom the mountain and surrounding lands were held in trust from 1826 to 1870, seems to have realized that the chief values of The Pilot were scenic. She had a trail constructed on the south side of the mountain and ladders on the side of the Knob. No further serious attempt to commercialize the mountain was made until 1929, at which time Mr. Spoon built wooden steps on the north side of the Big Pinnacle and constructed a toll road from the base of the mountain on the northeast to the summit of the Little Pinnacle.
>
> Fire occurred very regularly in the past and resulted in extensive burned-over areas. The most serious recent fire burned over most of the mountain in 1927, and practically all of the large trees are badly fire-scarred. Since the construction of the road, however, only two small fires have occurred.
>
> Summary: Quartzite caps the mountain. The soil is of a fine sandy loam and is derived from the underlying quartzite, a rock formation that is uncommon in the Piedmont. Lumbering, cultivation, disease, and insects have affected the vegetation, but the greatest disturbing influence has been that of fire.
>
> —Ruby M. Williams and H. J. Oosting (1944)[1]

1. Ruby M. Williams and H. J. Oosting, "The Vegetation of Pilot Mountain, North Carolina: A Community Analysis," *Bulletin of the Torrey Botanical Club 71, no. 1, sponsored by Duke University, Durham, North Carolina (January 1944)*: 23–45, https://www.jstor.org/stable/2481485.

Teresa Gillam moved to Atlanta, Georgia, and passed away April 30, 1877, a decade or so before the railroad came to Pinnacle, near Pilot Mountain. Almost as soon as the railroad came to Pinnacle, a train excursion combining visits to Pilot Mountain and a granite quarry in Mt. Airy became popular. Instead of dozens of climbers on the mountain from the various hotels that sprung up in the area, now hundreds of people were coming all at once. The Pinnacle train station was the gathering spot for an excursion on the mountain, but now they had a time window of only a few hours rather than an entire day. This train ride usually included a stop at the Mt. Airy granite quarry only ten or twelve miles from Pilot Mountain. The excursion stop for the quarry would last for an hour-long visit, followed by a Pilot Mountain excursion for four to five hours. Buggies, known as conveniences, would ferry people back and forth from the train station to the mountain. At the end of the mountain visit, the train returned to Greensboro by twilight. This excursion was in a similar style of a modern-day cruise, but returning to the train instead of the ship.

In 1896, the issue of the Gillam children and grandchildren's inheritance of the Pilot Mountain property came to a head. The issue was how they should distribute the trust, partly due to poorly-kept sales records of the property. Teresa's siblings were no longer involved in this part of the estate. For example, one sibling returned to New York City with his mother or step-mother. The land

FIGURE 5.3. Mt. Airy granite quarry with Pilot Mountain in the background in 2021. In the 1800s when it was simply a farm, this would have been a forty-acre granite bald spot feature, devoid of trees. W. L. Spoon has a wonderful story about that particular farm later in this chapter.

FIGURE 5.4. Ad in the The Greensboro Patriot Wednesday, May 8, 1895.

was part of a trust and the record keeping, sales of lands and timber, interest, and the fair distribution by the estate trustee were objected to repeatedly in court documents. The case, *Whitfield v. Boyd*, wasn't resolved until 1915. The case proceeding started at the Court House in Dobson, NC, on October 5, 1896.[2] This division of the Pilot Mountain property developed into the longest court case in North Carolina's history at that time, and it went to the NC Supreme Court in 1910. The long series of summons, complaints, amended complaints, answers, amended answers, evidence, exhibits of defendants, referee responses, judgement, and appeals with so many court documents it was later bound and stored in the State Archives of North Carolina in Raleigh. At the conclusion of the court case, Pilot Mountain was finally put up for auction. In the meantime, Pilot Mountain was still in the public eye during the trial.

2. Whitfield v. Boyd.

ANNUAL EASTER

EXCURSION

—TO—

MOUNT AIRY,

Granite Quarries and Pilot Mountain!

WILL RUN FROM

GREENSBORO

Monday, April 19th, 1897.

SCHEDULE AND FARE FOR ROUND TRIP:

Leave Greensboro 7 00 A. M., fare $1 00; Walker avenue 7.05, fare $1 00; West Market street 7.08, fare $1 00; Battle Ground 7 22, fare $1 00; Summerfield 7.36, fare $1.00; Stokesdale $7.52, fare $1.00; Walnut Cove 8.22, fare 80 cents; Germanton 8 40, fare 80 cents. Arrive Mt. Airy 10 10; Granite Quarry 10.25. Returning, leave Granite Quarry 11.15; arrive Pinnacle 12 20 P. M.; leave Pinnacle 5.50; arrive Greensboro 8.25. Children under twelve years of age half fare.

This schedule gives 50 minutes stop at the great Great Granite Quarries, and five hours and thirty minutes at Pilot Mountain. This will be the NICEST CHEAP TRIP offered this season. Take a day off and go with us on the

FIRST PICNIC EXCURSION

We will have first class service, prompt movement of train and run on schedule time. A trip to the Pilot Mountain has never failed to please. For further information see or write,

W. F. BOGART, Manager,

GREENSBORO, N. C.

FIGURE 5.5. Ad in The Greensboro Patriot Wednesday, April 7, 1897.

WHO IS W. L. SPOON?

If you visit Pilot Mountain today, you owe a debt of gratitude to William L. Spoon (1862–1942), who built the modern road up to the top of Pilot Mountain. Drivers going up and down the road should be watchful for hikers also on the road. Hikers have been walking alongside traffic up the road for nine decades now, even though hiking trails to the mountaintop are much more scenic.

W. L. Spoon became a civil and road engineer and a road advocate in Alamance County, NC. Born to George Monroe and Nancy Stafford Spoon, his father tragically died during the Civil War. His mother then married Michael Shoffner, a farmer and miller of Hartshorn, Alamance County, on whose farm Spoon spent his early youth.[3]

3. William L. Spoon Papers, 1858–1957, #04009, Wilson Library, University of North Carolina at Chapel Hill, https://finding-aids.lib.unc.edu/04009/.

FIGURE 5.6. Photograph of W. L. Spoon from unknown date. Courtesy of the W. L. Spoon Collection at the Wilson Library at UNC-CH.

Spoon began his education in the public schools of Coble Township, now a suburb of Greensboro, NC, seventy-six miles from Pilot Mountain. He spent one term at the Oakdale Academy and attended the Friendship School in Coble Township for two years, according to his biography at UNC-CH's Wilson Library. He began farming on his stepfather's land and taught in nearby public schools. In 1886, he entered a college preparatory course at Graham Normal College (later Elon College)[4] in Graham, NC. The following year, at the age of twenty-five, he entered the University of North Carolina, graduating in 1891 with BS and BE degrees. During his first year, he submitted plans for a road between Chapel Hill, NC, and Durham, NC, which were adopted and built. This was quite impressive for a farm boy in that era. After graduation, Spoon served as an engineer with the NC Geological Survey, supervising diamond drilling for coal and marble throughout the state (1891–1892). His love of travel

4. "Graham College," Historical Marker Database, last revised June 22, 2021, https://www.hmdb.org/m.asp?m=29894.

led him to spend a year surveying and producing a map of Alamance County, NC (1893–1894).

Following this venture, he worked in private roadwork, largely in Alamance and Orange Counties, NC. Not to be trapped in one state, in July 1895, Spoon taught school in the town of Moody, Texas. He visited North Carolina for the summer, and went back to Texas. Though when he was unable to find work there, he returned to North Carolina. There he operated as an independent engineer until June 1897, when he became a road superintendent and county engineer for Alamance County, NC, a position he held until January 1901.

Always able to find work somewhere, Spoon next traveled through North Carolina and Tennessee selling road machinery for the Good Roads Machinery Company of Kennett Square, PA (1901). He had become a road-building expert, which was useful when he bought Pilot Mountain. Remember, cars were new at this time and stimulated a greater demand for better roads. After staging road construction demonstrations in South Carolina, Georgia, and Alabama, and doing private engineering work, Spoon received an appointment as a special agent for the United States Department of Agriculture Office of Public Roads, an affiliation he maintained with various promotions until April 1909. After leaving the Department of Agriculture, he became a highway engineer for the NC Geological Survey and served as the highway engineer for Forsyth County, NC, from 1911 to 1913.

Spoon was clearly intelligent, resourceful, adventuresome, thoughtful, and inventive. He was often cited in local newspapers with having a new invention. He later returned to the Office of Public Roads as a senior highway engineer in 1913 and stayed until 1920. At age fifty-eight, he left federal service and became a partner in the firm of Spoon and Lewis, Consulting Engineers, in Greensboro, NC. He continued to receive assignments from the Office of Public Roads until 1931, when age regulations barred him from further service. He was sixty-nine years of age, fit, and active.

In his personal life, Spoon married Susan Adeline Vernon Neville (called Addie) in 1897 when he was thirty-five. They had two children: Nancy Miriam and William Mozart. In 1936, after twenty-nine years of marriage, Addie died. Four years later, in 1940, at age seventy-eight, Spoon married Ruth Baldwin. Spoon was prolific till the end, and a daughter, Willie Ruth, was born in 1943. Unfortunately, she never knew her father, as he died at age eighty-one shortly before her birth.

W. L. SPOON PURCHASES PILOT MOUNTAIN IN 1922

The Orinoco Supply Company of Winston-Salem, bought the mountain in 1915 at auction for the virgin timber that Andre Mathieu was so proud of during his ownership. The company might not have known of the long-standing history of fires around Easter, which also plagued them during their ownership bringing down their timber yield. During their ownership, there was a huge fire on the mountain much worse than the 2012 or 2021 fires. An account of fighting one of the annual Easter fires on Pilot Mountain a century ago was published in 1977 as a remembrance in *Our State* magazine:

> As my family watched that fire over a half-century ago, I could see the concern in my parents' faces. They were afraid that it might eventually burn down into the 'flatwoods' at the foot of the steeper slopes to envelope the homes of our neighbors. . . My father usually 'knocked off' from work about four in the afternoon, ate an early supper, and left home in time to reach the fire lane before dark. He took a pitchfork or a rake, a lantern, and plenty of chewing tobacco. Although the task was time consuming and a trifle dangerous, I'm sure that he enjoyed it as a change in his daily routine.
>
> When I grew older, I was privileged to go with him once, before the owners of the mountain began putting them out before they got well underway. I remember the excitement, a mixture of fear and anticipation of the unknown, that held me as we began the climb up the old logging road in the growing dusk. I felt my eyes burn in the smoke laden atmosphere and had to clear my throat occasionally of congestion. As we approached the lane, we heard the yells of the men already at the station.

FIGURE 5.7. This might have been an old wagon for tobacco or used by the Orinoco Supply Company to haul wood. This image was taken in 2022 on the Grassy Ridge and Mountain Trail Loop near the campground.

> Father let out stentorian roars in answer. At first I thought the yells sounded of distress, but I soon realized that they were just the means of the men letting others know that they were there. I could hear men from both north and south responding, thereby maintaining a sort of chain of communication around the mountain.
>
> —Zeb Denny (1977)[5]

Spoon had admired Pilot Mountain for years before he committed to buying it from the Orinoco Supply Company. Orinoco, despite fires, still had harvested virgin timber from Pilot Mountain in great quantities. The harvesting of timber continued during the private ownership years when Spoon owned the mountain. A lot of timber was cut down to clear the way for the new road up the mountain in 1929, as seen in the image at the start of this chapter.

> The park's master plan written at its founding says that it was the local custom to set fires on the mountain at Eastertime until the mid-1930s and that the last 'significant fire occurred in 1927. Two 'small' fires were reported between 1929 and 1944, but their extent, severity, and exact date were not recorded (Williams & Oosting, 1944). Between 1948 and 1969, there were two lightning-ignited and four human-ignited fires. The largest of these, in 1960 and 1961, were set by hunters and burned 1 and 5.7 ha (3 and 14 acres), respectively (Bell & Morse, 1970). The NC Park Service began using prescribed fire within the park in 2009 and since then the entire upland portion of the park has been burned at least once (Windsor, pers. comm.).
>
> —Dane Mitchell Kuppinger and Abigail Rich (2019)[6]

Spoon and his cousin and business partner, James E. Stafford, bought Pilot Mountain from the Orinoco Supply Company in 1922. They purchased 1,066 acres at $20 per acre with a total purchase price of $21,320. They did not remain partners long, as Spoon bought out Stafford's share in 1928. Being a road expert, it is no surprise that soon Spoon invested $13,000 and many hours into building the clay and sand road up the mountain in 1929. Spoon planned to earn back the investment in the road by charging per-person tolls in cars and walking. Perhaps his cousin did not share his enthusiasm for investing such a large sum of money in building a road.

---

5. Zeb Denny, "We Fought the Fires of Pilot Mountain" *The State*, May 1977.
6. Dane Mitchell Kuppinger and Abigail Rich, "Fire in Central Piedmont as Recorded by Fire Scars at Pilot Mountain State Park, NC," *Physical Geography* 41, no. 3, 2020, https://doi.org/10.1080/02723646.2019.1649008.

Spoon had admired Pilot Mountain all his life, but one particular drive by the mountain from Mt. Airy to Winston-Salem with his good friend George Hinshaw gave him the idea to buy it. At that time, the Pilot Mountain property was still involved in a court case well-covered by the media.

> On one occasion, returning from Mount Airy in company with our old friend George Hinshaw, as we passed Pinnacle and looked over at the Pilot, the remark was made that Winston-Salem should own that mountain as a city property and park for the masses of her citizens, and as a distinctive advertising means for Winston-Salem and her industries. Mr. Hinshaw observed at that time, with regrets, that the Mountain was then in the courts and could not be purchased. But in 1916 a court order was issued and the Mountain property was sold and bought by the Orinoco Supply people, who cut off and sawed into practically all of the original forest pine.
>
> —W. L. Spoon (1924)[7]

It is not clear what motivated Spoon and Stafford to buy Pilot Mountain, but two years later, after their purchase in 1924, they tried to sell it to Bailey Walker and Associates for $75,000—a profit of $53,700! This was a large sum of money for a two-year investment in those days. Even if Spoon had originally purchased the land to preserve it, that amount of money might have swayed most people. Walker is reported to have backed out because of a slump in the Florida real estate market, in which he was an investor. The formal reason had to do with a "defect" in the deed, despite the case being in the court system for almost two decades:

> However, the ostensible reason as given by Mr. Walker was a defect in the title due to a minor heir living in Georgia. The defect consisted of a failure of the court officials in Surry County to make proper records of the guardian appointed for this minor child, and hence as a matter of record it would appear that this legal requirement for the minor's protection had not been complied with. But, as a matter of fact, it was only a court record in Surry County at fault and as soon as the majority of the young man was attained, Judge Shepherd Bryan, of Atlanta, secured for me a quit-claim for this young man. So, now no shadow or doubt exists in the title of the property.
>
> —W. L. Spoon (1924)[8]

7. William L. Spoon Papers, individual letters folder.
8. William L. Spoon Papers, Box 10, File 436.

BUILDING MATERIAL

Sash, doors, blinds, flôoring, ceiling, siding, lime cement, plaster, shingles. Write or phone us for quotation. We will save you money.

ORINOCO SUPPLY COMPANY

Cor. Main and Second Phone 362 or 803

FIGURE 5.8. Ad placed in the Winston-Salem Journal for the Orinoco Supply Company in 1915, which was the year they bought Pilot Mountain for timber.

SPOON'S EARLY OWNERSHIP: EASTER SERVICE ON PILOT MOUNTAIN

Easter services on Pilot Mountain had been going on for decades, so it might not have been a surprise to Spoon when he got a letter from a minister dated March 11, 1924, asking to have a service on the mountain with a congregation. Spoon responded,

> Your letter of the 11th relative to the privilege you ask to hold (of holding) religious services Easter Sunday on Pilot Mountain is acknowledged.
>
> We do not object to responsible religious organizations holding services on Pilot Mountain provided it is conducted reverently and the gathering is sober and orderly and do not mutilate and damage the trees or other property. The spirit of true worship of God is welcome.
>
> Hope you may be filled with the inspiration that noble specimen of God's handiwork ever suggests to you and all men – to stand upright and be a beacon of light, a Pilot to men.
>
> Very truly yours,
>
> —W. L. Spoon (1924)[9]

9. William L. Spoon Papers, correspondence papers, Box 9, File 426.

SURVEY AND SPOON'S ROAD TO THE TOP

There is mention of an early bridle path for horses by a Reverend W. E. Poovey in a lone newspaper article, found in the Joe Matthews Collection titled "Pilot Mountain, Landmark of the Piedmont."

> Some pioneer settler in the vicinity of the Pilot, using no instruments but a practiced eye and an ax blazed a bridle path from the base to the summit. This was later widened to admit horse-drawn vehicles. But now a well graded automobile road wends its way along the western slope to the secondary pinnacle, and parking space is available for hundreds of cars only a few steps from the base of sturdy winding stairways that take the stout-hearted to the top of the two-hundred-foot stone table land. A nominal toll-charge is made to cover the cost of keeping the graveled roadway in good order the year round.
>
> —Rev. W. E. Poovey (unknown date)[10]

This report of a way up the mountain from the north side by horse, and later horse drawn vehicles, is the only mention found anywhere. Dozens of newspaper accounts in newspapers talk about going up the south side by horse to the mineral spring. That only got you halfway up that side of the mountain; the rest of the way was by foot from there to the top. This combination of transportation by horse and foot meant it took at least an hour and a half to two hours to reach the top portion of Pilot Mountain. By nearly all accounts, this mountain hike was exhausting.

Did both the untrained earlier pioneer, with his ax, and professional road builder Spoon, with his modern survey equipment, discover the same ancient roadbed from some prehistoric time period or civilization or both? The roadbed may have acted as a sort of blueprint right in the land for an original survey of the road. The entire first chapter of this book reveals how quickly nature can reclaim something man-made in just one lifetime on this mountain. How old the road up Pilot Mountain might be is anybody's guess, given this lone account by Poovey. If this is a road originally built by an ancient civilization, any relics are now well preserved under it. The mountain is quite resilient and able to return to its pristine fauna in a surprisingly short period of time. Compare this to the recovery of the land after the Mount St. Helens volcanic eruption, which took a decade instead of the predicted thousands of years.

10. W. E. Poovey, "Pilot Mountain, Landmark of the Piedmont," n.d., Mount Airy Museum of Regional History. Context is given by a letter written to Spoon by Poovey on July 13, 1931, found in in Box 9, File 427, Mount Airy Museum of Regional History.

GRADE OF THE PILOT MOUNTAIN ROAD

"The road enters the mountain park on the east side, swings up westward—then turns northward in all about 1,500 feet to the toll gate located on old trail," W. L. Spoon explained in his 1931 ten-cent pamphlet "The Story and Facts of Pilot Mountain."

> The steepest grade is encountered 100 yards before reaching the gates. It is 16%. Cars able to make this grade can make all the rest to the top. At the gate is plenty of water for cars—and no trouble to get it from a hose for free. From the toll gate the road runs northward and turns westward to the extreme west end; there makes a 'Hair-pin' turn back eastward, about 2,000 feet more you are in the park. On this section you will see a huge arrow head shaped rock setting upright on your right. This rock was taken out of the road by the steam shovel operator and set up where it now stands. It was named 'Harrison's Signal Rock.' Just after passing it you come suddenly out in view of the great pinnacle. Another safe hairpin turn and you are in the upper park. Drive to the summit of the loop and park your car.
>
> —W. L. Spoon (1931)[11]

FIGURE 5.9. This is the actual road grader W. L. Spoon used for the road up Pilot Mountain. The road grader is now owned by the family of a childhood friend of Andy Griffith, Earlie Gilley. Someone had to drive the grader, pulled by a horse or two, and another had to add weight to the back end by sitting on it.

11. William L. Spoon Papers, correspondence letters, Box 9, File 426.

JULY 4, 1929, OPENING THE ROAD

Thousands of people will question at once: What is the road like? Can one drive an automobile all the way to the top of the mountain? The answer is that the road is not a trail, at all, but a highway which will accommodate as many autoists as care to go up or come down. Starting a short distance west of the town of Pinnacle, the new mountain road leaves State Highway 66 and winds its way about a mile and a half to the foot of the mountain. From that point it is two miles to the top. After the fashion of mountain roads, this driveway winds its way up the heights among virgin forest, hardly ever penetrated by man. The highway starts to ascend from the eastern end. Gradually it works its way along the steep slope of the mountain – a slope as steep as stone and earth can be made to lie without sliding and rolling – until it reaches the western end. Here it reverses itself and runs eastward again finally coming to an end in short graceful loops at the absolute summit of the little pinnacle.

From the starting point of the road to the summit of the mountain the grade is ten per cent, with one or two exceptions where a slightly steeper ascent is made for a few feet. Therefore, it is possible to drive the entire distance in high gear or second. The land furnished excellent road-building material, a soil for the greater part made up of coarse gravel. Only a single stretch of some dozens of yards did the engineer find it necessary to haul gravel from elsewhere to complete the road.

—Harvey Dinkins (1929)[12]

Was the gravel bed that was found during the road's construction natural or man-made? Gravel is an underground feature of the early Mound Culture across the entire eastern United States. The Mound Culture will be addressed in book two of this series. Regardless, the W. L. Spoon Road was opened on July 4th, 1929, and was a huge success. Automobiles that went up the new road in 1929 were charged 50 cents for each passenger over the age of 16 and 25 cents for children between the ages of 10 and 16. Walkers were charged 25 cents.[13] There were also discounts for special parties (e.g., schools, civic organizations). There was a public relations blitz in newspapers when the brand-new clay and sand road up Pilot Mountain opened on the fourth. It is noted in the W. L. Spoon archives that the gate receipts for that single opening day was $500 at the tail end of the Roaring Twenties.

Spoon's gamble on the mountain road furnished a reliable income its initial years of operation. In July 1931, the Pilot collected a toll gate of

12. Dinkins, "Dreams Come True."
13. Mark Farnsworth, "Sentinel of the Piedmont: A History of the Mountain, 1700–1950" (University of North Carolina Greensboro, master's thesis, 1996), 48.

FIGURE 5.10. The era of the steep stairs replaces nearly a century of the use of ladders to climb the Big Pinnacle. The ladders, stairs, and later even a flagpole, were often victims of lightning strikes over the past two centuries. Courtesy of the Wilson Library at UNC-CH and the W. L. Spoon Collection.

> $355.50. The year 1933 generated a yearly gate receipt of approximately $1368. By 1938, the yearly income from the tolls had fallen to $523.20. The Great Depression appeared to be having a detrimental impact on visitation to Pilot Mountain.
>
> —Mark Farnsworth (1996)[14]

This barely made a dent in the $13,000 road-building investment, not to mention the upkeep costs and annual taxes.

14. Farnsworth, "Sentinel of the Piedmont," 51.

> In 1929 I made a survey and built a road from the highway to the top of the little pinnacle which cost me in money and time approximately $13,000.00. The property is now valued on Surry County tax books at $16,000.00 for purposes of taxation. Since the road was built an average of approximately five thousand people have visited the Mountain annually, many of whom do not go up and a large percentage of those who do go up walking, making the revenue derived from the Mountain as it now is enough to take care of the road but not sufficient to render a substantial profit.
>
> —W. L. Spoon (1939)[15]

In an interview, Spoon told Harvey Dinkins about the new road:

> I wanted to make it possible for thousands of people throughout the country to see this great natural wonder who otherwise could never have done so, you know, very few people could climb the mountain before the road was built. Now, even the invalid can ride to the top in ease and comfort and can see as much as the sturdiest climber. If I have helped anyone by this, I am glad. At any rate, it was only the carrying out of a dream.
>
> —Harvey Dinkins (1929)[16]

Dinkins's newspaper account on July 14, 1929, also reported that, in the "Boulder Cove" area on the road, about three-fourths of a mile from the top, black bear tracks had been seen several times around the mineral spring.[17] That was where automobiles might have stopped to refresh their overheated radiators, and walkers found the north side mineral spring just as refreshing as the south side one. Bears were so common that it wasn't a big deal. Actually, the timber rattlesnakes were more of a concern on Pilot Mountain, especially around the base of the mountain near the apple orchard, or on the very top of Big Pinnacle.

MANY IMPROVEMENTS FOR PILOT MOUNTAIN IN 1929

In addition to the newly surveyed road Spoon built, once he had the cars parked on the top, he needed a new path across the saddle area. Harvey Dinkins wrote in the *Winston-Salem Journal and Sentinel* on Sunday, July 14, 1929, ten days after the opening of the road:

15. William L. Spoon Papers, individual letters folder.
16. Dinkins, "Dreams Come True."
17. Dinkins, "Dreams Come True."

> Possibly the new trail connecting the two pinnacles and the stairway by which the tourist is able to ascend the big pinnacle offer greater interest to the prospective mountain climber than any other feature encountered at the top of the mountain. From the park at the top of the little pinnacle the new trail winds downward a short distance and crosses by an easy approach to the great isolated mass of stone known as the big pinnacle. It shortens the distance by many hundreds of yards and is far easier. Lovers may go the other way about, if they like!
>
> The stairway by which one ascends the big pinnacle is located in the exact spot where the old ladders stood, which place offered the easiest ascent. By 106 steps, broken in three places by landings where rests may be taken, the naturalist now reaches the top where, in other days, a perilous climb up shaky ladders made even the person of steel nerves shudder.
>
> —Harvey Dinkins (1929)[18]

W. L. SPOON WAS THE FIRST TO WANT AN AIRFIELD ON PILOT MOUNTAIN

Spoon in 1929, and the next owner fifteen years later, J. W. Beasley, both wanted an airfield to land planes, gliders, balloons, and helicopters on Pilot Mountain. What was the motivation for these two owners of Pilot Mountain? It was certainly a novel idea in the time of early air travel. Maybe it would have been a tourist draw to watch planes land on the mountain. Since the mountain is so far above sea level, this could have made it among the highest airfields, if not the highest, east of the Rockies in the United States. Spoon first mentioned the airfield in his ten-cent pamphlet created for the grand opening of the road July 4, 1929. He said he had cleared the land and was about to create a grass field for landing planes on top of Pilot Mountain. This was the time period after the Spanish Flu pandemic and the Roaring Twenties, which followed with money freely flowing everywhere. Big dreams like the airfield had to be put on the back burner for Spoon's Pilot Mountain as the world entered the Great Depression.

> About 300 yards west of the park is the landing field in process of development. When completed and sodded it will enable airplane(s) to land on the mountain top. As a glider spot point, it has no equal in America.
>
> —W. L. Spoon (1929)[19]

18. Dinkins, "Dreams Come True."
19. W. L. Spoon, "Pilot Mountain: A Gem of Scenic Beauty," 1929, W. L. Spoon Archives, Wilson Library, University of North Carolina-Chapel Hill.

In a later undated but updated version of the "Gem" pamphlet, when describing the gently sloping mountaintop where the picnic area is now, Spoon stated, "Small planes and helicopters have landed here."

> Probably, the most creative idea suggested by Spoon involved the opening of a glider landing strip on the little pinnacle. Spoon landscaped an area on the little pinnacle, which provided a landing strip for glider airplanes. He felt confident that it would prove a great asset to the interesting sport of glider aircraft.
>
> —Mark Farnsworth (1996)[20]

Repeated inquiries to the Federal Aviation Administration to find out if an airfield application started by Spoon or Beasley resulted in their Memphis Office saying by email and phone that they could find no application in their digital files. At the time of publication of the book they only had access to the digital files.

CHARITY ON PILOT MOUNTAIN

The Mills Home sent a truckload of their children to Pilot Mountain for a picnic. The following appeared in "Charity and Children" for March 19, 1931, that Spoon reprinted in a pamphlet sold at the ticket gate:

> Mr. McKoin and Mr. Poplin took the truck loaded with their Sunday school classes of intermediate boys Saturday morning and went to Pilot Mountain for the day. The matrons had prepared a basket filled with the kind of food boys need after a mountain climb.
>
> When they passed through Winston-Salem, Hine and Gore, wholesale produce merchants, good friends of Mills Home, donated a bunch of ripe bananas. It was a tremendous bunch of bananas but there were only a few on the stem when the boys got home. Pilot Mountain is owned by W. L. Spoon, of Burlington. He has had a road surveyed to the very top of the mountain. The truck had no difficulty at all in making the climb. After reaching the top of the mountain they parked the truck right by the pinnacle and went up to the top by means of a good and safe stairway. There is a nominal charge for the use of the road but the view is worth many times what it costs. Mr. N. G. Amick is the very courteous keeper of the premises. Mr. McKoin and Mr. Poplin were perfectly willing to pay for the privilege of going up but were not allowed. Mr. Amick said they

20. Farnsworth, "Sentinel of the Piedmont," 52.

wanted to have a part in the boys' pleasure. Few people realize what a wonderful view can be had so near here – only forty-five miles from Thomasville. They recommend the trip and the view without reservation.

—Staff Writer (1931)[21]

Spoon also thought about making Pilot Mountain a bigger tourist destination. At this time, his archives show an estimate for an elevator tower to the top of Pilot Mountain from the Westbrook Elevator Company. It was going to be run by a gas generator, but the idea never got past the planning stages, possibly due to the economic crash in 1929. The shift from the Roaring Twenties to the Depression thwarted many of his grand ideas for the mountain.

With the potential for the glider airfield atop Pilot Mountain, Spoon must have decided that when it was not in use it could be used as a parking lot for up to five hundred cars. Back then, you could park as close to the cliff on the south side as you dared. The open parking area had oak and chestnut shade trees around the perimeter. For a time, tobacco was grown in this part of the mountain where the picnic area is now. Over eight acres were tended at the bottom of the mountain, and one of those fields is where the new Pilot Mountain Visitor Center, built in 2021, now stands.

FIGURE 5.11. One of the hairpin turns on the clay and sand road built by W. L. Spoon in 1929. On the back of the original photograph, in pencil, it notes this is the hairpin turn on the west end. Courtesy of the W. L. Spoon Collection and the Wilson Library at UNC-CH campus.

21. W. L. Spoon, "Pilot Mountain: A Gem of Scenic Beauty," 27.

WE CALLED IT "GOING TO THE MOUNTAIN"

My earliest memories of going up there is everybody would get into granddad's car [W. L. Spoon]. He always had a two-seater black Ford car. I remember being tucked in the back seat of that with his heavy lap robe thing that was huge. They must have had it for buggies and things like that. It was really heavy but it was really warm. My mother would tuck us into that. I remember these trips because we would call it 'going to the mountain.' We got up there and we drove all the way up to the big dome. Then we climbed up those steps that he built to the very top. Until I was grown 'going to the mountain,' to me, always meant going up to the top of the big dome steps. After he died, when it was sold, the decision was made to tear them down because they weren't considered safe. They were safe to me. . . I do remember being up there and there were rattlesnakes. I never saw one, but in 1943 when I was ten years old, for reasons of not having clean water to drink where we lived in Alamance County, we had to live elsewhere. They were afraid the spring where we lived was contaminated with typhoid. They did not want us to stay there. So, I lived in High Point with my aunt and uncle. My sister lived up on Pilot Mountain with Uncle Richard and Aunt Lois.

By that time my grandfather had died. He died without a will so they all had fun carving up his estate. So, while all that was going on Uncle Richard and Aunt Lois lived up on Pilot Mountain. Uncle Richard ran the gate. He collected money from people that went up that way. There was still money to be made that way. My sister Nancy lived up there with them and I was so jealous because I wanted to live on Pilot Mountain. They would send that little girl into the apple orchard and there were rattlesnakes in there. Oh, and I now think 'thank God I was in High Point.

—Frances Alexander Campbell (2020)[22]

THE GREAT DEPRESSION CLEARLY HURT PILOT MOUNTAIN REVENUES

The strains of the Great Depression made Spoon's attraction seem less and less desirable. Spoon informed Joseph Hyde Pratt that while visiting the mountain one Sunday that 'quite a large number of people came up but more than half turned back at the gate.

—Mark Farnsworth (1996)[23]

22. Campbell, phone interview.
23. Farnsworth, "Sentinel of the Piedmont," 51.

In 1939, as W. L. Spoon aged, he attempted to sell the mountain. He placed a "For Sale" sign on the highway that said, "Pilot Mountain for Sale, 1066 Acres, see W. L. Spoon – 432 Jefferson Bldg., Greensboro, N.C., Dial 4930." You'll remember that Spoon bought Pilot Mountain in 1922 for $20 per acre. Almost two decades later, at the tail end of the Great Depression, Spoon was trying to sell it for around $30 per acre. The deed showed 1,066 acres in the tract to be sold for a total of $32,000 as the intact asking price. While this got a lot of interest by tourists passing by, there were no takers.[24]

> In driving over the highways of the state you frequently see signs of 'House for Sale,' 'Fresh Honey for Sale,' 'This Farm for Sale,' 'Chicken and eggs for Sale,' but we're willing to bet that never have you seen a sign – "Mountain for Sale" – except in one place.
>
> We were driving from Winston-Salem to Mount Airy last Wednesday and it wasn't long before Pilot Mountain swung into view in all of its unique grandeur. We suddenly realized that we never had been atop of Pilot, so when we reached the town of Pinnacle we swung off to the left and headed for the mountain.
>
> At the junction of the road with the highway, we observed the 'Mountain for Sale' sign, and we paused to take a picture of it.
>
> —Carl G. (1940)[25]

Having no takers, Spoon had to find other ways to make the mountain pay for itself. The very astute Spoon figured out that feature article space was being bought from the magazines and newspapers of the time. At first, Spoon complained about having to buy ad space for events on Pilot Mountain.

> Aside from the Winston-Salem Journal and a short note in the Greensboro Daily News no paper in the state publicized it in the least without pay. Yet full page write ups have appeared of Morrow Mountain, Chimney Rock, The Sauratown State Park, etc. I have been puzzled to know the Psychology causing this attitude. So far as I know Chimney Rock is wholly a private development, but Morrows Mountain and Sauratown are wholly from public funds or nearly so. True the lands for the park, in each case, have been provided by private individuals, I take it. Yet when it comes to scenic beauty and sweep of the eye Pilot Mountain out distances them all. It is unique in outline and geological wonder. It stands apart from its sister Sauratown Mountains as a sentinel; the Indians

24. William L. Spoon Papers, Box 10, File 437.
25. Carl G., "Mountain for Sale," *Our State*, April 13, 1940.

> called it, 'Jomeokee,' which interpreted means the 'Great Guide.' Standing west of all its sisters and often climbed by Daniel Boon(e) in his early manhood he found a happy hunting ground for his eye and imagination.
>
> —W. L. Spoon (1939)[26]

April 9, 1940, in letters to the editor of The State, taken from the W. L. Spoon Archives at Wilson Library at UNC-CH, when Spoon was seventy-nine years of age:

> I love that old Mountain and am selling it because of my age and financial inability to do for it what it deserves. I do not care to sell it to a man without financial resources to take care of it as a park and preserve its beauty and make it available to all who are open to an inspiration from the majesty displayed. To all others it is just a rock and nothing more to them.
>
> —W. L. Spoon (1940)[27]

ANOTHER CHANGE OF MIND

> Dear Dr. Pratt,
>
> Have been seriously considering the proposition of making a State Park out of the mountain since my visit with you. I was up at the mountain Sunday and quite a large number of people came up but more than half turned back at the gate. They insist it should be free. Of course under present conditions that is unsound and unjust reasoning because of the fact that there is no other way to maintain the road, protect it from fires and otherwise police it and maintain order beside charging a toll. I have found to my great surprise the strong opposition to any sort of a toll charge. People are looking for a free park and there is no way of providing it except by making it a State Park and let the entire state take care of it and make of it a Free Public Park. As it is I have to pay heavy taxes and give over half of the receipts to protect it and the people.
>
> Then too, the accessibility, the uniqueness of the whole mountain eminently marks it as an object of great interest to the average man as

26. W. L. Spoon, letter to Sanford Martin, editor, *Winston-Salem Journal*, August 26, 1939, Box 10, File 436, W. L. Spoon Archives, Wilson Library, University of North Carolina-Chapel Hill.
27. W. L. Spoon, private letter to the editor of *The State* magazine, April 9, 1940, Box 10, File 437, W. L. Spoon Archives, Wilson Library, University of North Carolina-Chapel Hill.

well as the most advanced scientist. It stands out as a major geological wonder. The view from the top is a perfect Cyclorama! From the Park Highway, at many points, it stands out like a great ship riding on the eastern horizon. No other single mountain peak, in the state or out of it, can compare with it in supreme uniqueness, you are entirely sound in your judgment when you decide the state should own it and not the federal government. The property is too small to meet the requirements of the United States park specifications but quite large enough for the state to develop a perfect play and recreation park. There are two large springs on the northside of perfectly clear pure water issuing from the solid rock. This supply is abundant for all park purposes. Beside these two large springs there are several small springs. Notably the old trail spring on the east and the cliff spring on the west, taking it all in all it is the most perfectly self contained unit, standing out as a Great Monument to the creator – an index finger to higher and better things above, an inspiration to every normal mind.

I want to see it a state park and it should be started at once.

Very sincerely yours,

—W. L. Spoon (1940)
Cc Dr. Ronthaler[28]

Economic pressures surely had a hand in Spoon's change of mind. In 1940, after repeated attempts to sell the land and make a profit, Spoon wrote, "I want to see it a state park and it should be started at once," to Dr. Joseph Hyde Pratt of Chapel Hill, NC on April 30, 1940.[29] It would be another owner and twenty-eight years later before the state park idea took hold and became a reality.

GEOLOGY OF PILOT MOUNTAIN ACCORDING TO W. L. SPOON

In response to a letter from Mr. J. Drew Martin, III, of Columbus, Ohio, Spoon recalled the geology of the mountain and the area. Martin had inquired about buying Pilot Mountain after driving by the mountain and seeing the unusual "For Sale" sign.

I walked over quite a bit of the Mountain yesterday and the more I view the property and realize its advantages, the more certain I am that I am

28. W. L. Spoon, letter to Joseph Hyde Pratt, April 30, 1940, Box 10, File 437, W. L. Spoon Archives, Wilson Library, University of North Carolina-Chapel Hill.
29. Spoon, letter to Pratt.

parting with the most valuable property that it has ever been my privilege to own. Potentially and actually, it has a value far in excess of the price I have named. The Mountain is a most wonderful geological formation and has stone of every grade of hardness from flexible sand stone to crystalline rock as hard as French buhr. The stone on the Mountain is so laminated that is quarried easily into slabs of almost any dimensions and polished almost like marble. I have never wanted to convert Pilot Mountain into a quarry proposition; nevertheless, its potential value in this direction is much more than perhaps you would think.

Perhaps you may know or have heard the story of the Mt. Airy Granite Quarry which, many years ago, was bought as a farm. During the trade the purchaser [John Gilmer in 1872] objected to the 40 acres of stone then exposed as worthless and the seller agreed to reserve it and eliminate it from the price of the farm on an acreage basis. As a result of this, Mt. Airy Quarry is known practically all over the world but the farm is utterly forgotten.

Similarly, Pilot Mountain has potentialities as a stone quarry far greater than anyone today realizes, but I have endeavored to avoid the quarry features and rather preserve its park beauty and value. And so, I am selling it as a park proposition and at a price far below the farm prices for similar mountain land adjacent it.

FIGURE 5.12. On the back of this W. L. Spoon-era photograph, it states this is the view just before crossing the Grindstone Ridge. The effects of the fires of the 1920s are visible in this image. The notch on the Big Pinnacle, which might hold archeoastronomy secrets, is seen here—just like it was revealed after the fires in 2021. Courtesy of the W. L. Spoon Collection and the Wilson Library at UNC-CH campus.

> From the geologist's point-of-view this is a unique mountain of the world. Nothing like it to (be) found elsewhere in the world. It's a challenge to any geologist to clearly and satisfactorily outline its history and satisfactorily account for all the strange conglomerations of schist, laminated and stratified sandstone and granite shot through with massive quartzite intrusion, all folded and twisted into a metamorphic mass, in places, without the slightest volcanic evidence. It is simply an indefinable display of a stupendous power in the earth's court.
>
> —W. L. Spoon (1939)[30]

The following is taken from a display at the Mt. Airy Historical Museum regarding the Mt. Airy granite quarry. This is included in the book since they used to have day-long train excursions to both the quarry and Pilot Mountain in the 1890s. In addition, this is background about the granite bald spots spread out across the countryside of this landscape, which is home to the Nunne'hi mentioned so often in the Cherokee Oral Tradition. The Nunne'hi are discussed in fascinating detail by the Cherokee in the last chapter of the second book of this series:

> The Mount Airy quarry, largest open-faced granite quarry in the world, is in the shape of an oyster shell and covers about ninety acres of operation. In 1889, Thomas Woodruff and Sons, railway station builders of Greensboro, recognized the potential value of the granite, originally thought to be just flat rock. Today, the North Carolina Granite Corporation employs several hundred people, some of whom are second and third generation employees.
>
> Mount Airy granite has its origin as molten magma within the earth. Over many centuries, contractions of the earth's crust thrust the magma to the earth's surface. There it cooled and solidified to form the stone we know as granite. Because Mount Airy granite is a homogeneous mass (solid mass), free from natural bed planes and vertical cracks, stone quarried today matches perfectly that quarried 100 years ago or 100 years from now!
>
> Using up-to-date equipment and technology, the North Carolina Granite Corporation produces from the biotite granite a product small enough to feed canaries and large enough to construct buildings. The granite in these products has three main components: feldspar, quartz, and mica. Quarrying the deposits for the diverse products is accomplished by a process called lifting: heat expansion in the summer and cool contraction

30. W. L. Spoon, "The Story and Facts about Pilot Mountain in Surry County, NC," 1939, image folder, File 4009, W. L. Spoon Archives, Wilson Library, University of North Carolina-Chapel Hill.

in the winter, allow for stone extraction from the quarry. For over 100 years, the quarry at Mount Airy has been in full operation. Geologist indicate the quarry can be worked for another 500 years without exhausting the supply of granite. The Amoco Building, Chicago, Illinois, The Tallest Granite Clad Building in the World.

—Mt. Airy Historical Museum display (2021)[31]

MORE ATTEMPTS TO CREATE A CITY PARK

About ten months ago the Smithdeal Reality and Insurance Company asked me for an option on Pilot Mountain and I gave them a net price to me of $30.00 an acre, which in round figures is $32,000.00 for 1066 acres called for in the deed to Pilot Mountain that I hold. As a matter of fact, the acreage is considerably more than that but for our purpose the deed acreage is satisfactory. This firm contacted, and was turned down in every instance as in no way interested in the matter of making Pilot Mountain a park for the City of Winston-Salem and owned by the city.

—W. L. Spoon (1940)[32]

FIGURE 5.13. Another view of the North Carolina Granite Corporation quarry from over Mt. Airy, North Carolina, in 2022.

31. "North Carolina Granite Quarry," museum display, Mount Airy Historical Museum.
32. W. L. Spoon, letter to Sanford Martin, editor, *Winston-Salem Journal*, 1940, Box 10, File 437, W. L. Spoon Archives, Wilson Library, University of North Carolina-Chapel Hill.

R. J. REYNOLDS INVOLVEMENT REQUESTED IN WRITING

Spoon's "For Sale" sign got some written inquires, but the mountain didn't sell until J. W. Beasley bought it in 1944 at auction from Spoon's estate after he passed. Spoon approached R. J. Reynolds to buy the mountain in 1940, but Reynold's attorney responded within twenty-four hours to decline the offer.[33]

END OF AN ERA OF PILOT MOUNTAIN GROWTH AMID HARDSHIP

W. L. Spoon passed away on August 28, 1942, at the age of eighty-one, and Pilot Mountain became part of his estate. Spoon was the biological father of three children, and he also had four foster children. Without a will, the Spoon estate was handled by probate court and eventually the mountain was put up for auction again. An auction was held three different times before its sale was finally announced in early 1944.

FIGURE 5.14. Courtesy of the Mt. Airy Historical Museum.

33. R. J. Reynolds, letter to W. L. Spoon, Box 10, File 437, W. L. Spoon Archives, Wilson Library, University of North Carolina-Chapel Hill.

FIGURE 6.1. Stone face on a Pilot Mountain State Park trail that you walk under.

6

J. W. Beasley and Pearle Beasley Modern Ownership in 1944

FIGURES 6.2 and 6.3. Painting of J. W. Beasley and photo of Pearle Beasley from their archives with permission from granddaughter Patricia Sperry and grandson Dr. Gene Glace.

J. W. Beasley (a.k.a. John W. Beasley) was a native of Patrick County, VA, who always dreamed of buying a mountain.[1] Early in his life, he rode a horse down from Virginia to Ararat, NC. He was then known as a "tinker," one who fixed things. First, he tinkered on small things, and later by buying, fixing, and selling farms.[2] It was on one of his horseback trips when he first met his future wife, Pearle Forkner, who lived in Ararat, NC.

Beasley was educated in Stuart, VA, and then lived briefly in Richmond, VA, before moving to Pilot Mountain in 1913. Beasley owned a barbershop in downtown Mt. Airy, got into the oil business, sold clothing for a time, and was a young

1. Shore, phone interview.
2. Shore, phone interview.

FIGURE 6.4. Postcard of the J. W. Beasley City Barber Shop on South Main Street in Mount Airy, North Carolina. Courtesy of Patricia Sperry.

editor of a newspaper called The Pilot Mountain Citizen.[3] In 1916, the offices of The Pilot Mountain Citizen, run by O. N. Swanson, were destroyed by fire. The building was owned by Beasley, the editor of the paper.[4] Beasley opened a Ford dealership in Richmond, VA, and then a Chevrolet automobile dealership in Pilot Mountain.[5] While living in the town of Pilot Mountain, he served as a town commissioner, was a member of the Masonic Lodge number 493, and was a member of the school board.[6] This life of service and enterprise is a hallmark of all the Pilot Mountain owners all the way back to Andre Mathieu.

BEASLEY EARLY-MORNING WEDDING

During this era, the train schedule had an impact on social events such as weddings. Most weddings were held early in the morning around Pilot Mountain in order for the newly wedded couple to catch the morning trains out of town. Such was the case of John [a.k.a. J. W.] and Pearle Forkner Beasley who were married on 7 September 1910 at the home of the bride's mother in Ararat, NC. The nuptials took place at seven o'clock in the morning followed by a wedding breakfast. The festivities were all concluded prior to the departure of the morning train. The newlyweds traveled to Washington, D.C., on their wedding trip and honeymoon.

3. Staff writer, *The Charlotte Observer*, June 1, 1916; Staff writer, *The Western Sentinel*, September 15, 1914; Staff writer, "Assistant Editor Pilot Mountain Paper," *The Western Sentinel*, June 3, 1916.
4. Staff writer, "Firt [sic] at Pilot Mountain Friday," *The Cleveland Star*, March 30, 1917.
5. Staff writer, "Beasley Chevrolet Company," *Rocky Mount Telegram*, October 28, 1948.
6. Shore, phone interview.

A visit to friends and neighbors was not a casual drop-in, especially for people living in the surrounding countryside or another town. Usually, the distance was such that visits required the entire day. The visiting family would pack a picnic lunch and leave home early in the morning, spend what remained of the day with friends or family, and return home before nightfall.

—Carolyn Boyles (2012)[7]

BEASLEY FAMILY IN THE H. LEE WATERS' FILM IN 1939

H. Lee Waters traveled the state, filming as many North Carolinian locals as he could find, which often included groups of a particular town during the Depression Era.[8] A few weeks later, he returned for a film's grand opening in that same town and was received with much fanfare. People who knew they were in the film were curious and came to the first showing. Then, they would call their friends to come down to the theater to see the following shows, which sold out in a rush.

FIGURE 6.5. J. W. and Pearle Beasley. Courtesy of Patricia Sperry.

7. Carolyn Boyles, *Early Days of Pilot Mountain: A History and Genealogy*, (Chelli & Associates, 2012), 11–12.
8. H. Lee Waters, *Pilot Mountain (N.C.), circa 1939 (Reels 1–2)*, H. Lee Waters Film Collection, David M. Rubenstein Rare Book & Manuscript Library, Duke University Libraries, https://repository.duke.edu/dc/hleewaters/RL10075-DBCAM-0022_01. https://repository.duke.edu/dc/hleewaters/RL10075-DBCAM-0022_02.

Rather brilliant marketing at a time when money was so tight. Waters did not miss the Town of Pilot Mountain in his travels throughout the state. Part of Waters' film of Pilot Mountain features the John W. Beasley home (owners of Pilot Mountain from 1944 to 1968), Pearle Beasley's Garden, and some of the Beasley family riding horses in town.

Tickets to see the film on May 15 and 16, 1939, at the Pilot Theater cost $0.25 for adults and $0.10 for children. 421 adults and 108 children paid a total of $116.05 for four showings of the film on a Monday and Tuesday. Waters split the gate 50-50 with the theater owner and got $22 in advertising, according to page 128 of Waters' record books. Waters's half was $58 minus $22, yielding $36. Minus his film, equipment costs, and his time, he didn't net a lot, but any money made in the Depression was a lot, and it gave him something fun to do. He used 760 feet of film for the "Town of Pilot Mountain" film. Many of Waters's original films are housed at Duke University, and many are available on YouTube, including the Pilot Mountain film.

PILOT MOUNTAIN STATE PARK ATTEMPT IN 1944 FAILED

After Spoon passed away on August 28, 1942, the mountain was up for auction again as it had been in 1915. Politics and money stood in the way of the people of the state of North Carolina buying Pilot Mountain to become a state park this time. It would be another twenty-two years before that dream would come true.

FIGURE 6.6. View of the Ledge Spring Trail, Little Pinnacle, and Big Pinnacle from the south side of Pilot Mountain, left to right. The burn marks from the 2012 fire just below the Big Pinnacle are still visible in 2022, seen here.

Pilot. For several months there has been a movement on foot to encourage the purchase by the State of Pilot Mountain, one of North Carolina's most widely known landmarks. More than 1,000 acres of land situated at the top of the mountain and including the 'Knob' became available last fall at the closing out of the estate of W. L. Spoon. Citizens of Winston-Salem notified the Department of Conservation and Development that the land was available and suggested the State purchase the land. The department turned over the suggestion to Governor Broughton.

Funds. Although citizens of Winston-Salem and other points in the vicinity of Pilot Mountain (the mountain, in Surry County, is about 35 [20] miles from Winston-Salem) are anxious to see the landmark fall into State hands, the Governor has expressed the opinion that the Department of Conservation and Development cannot make the deal, since the department has no funds available for such a purchase... The Governor is of the opinion that legislative action is necessary to make the funds available for purchase of the land... Those who favor State purchase of the mountain contend that public ownership of Pilot Mountain would insure preservation of the landmark and also would be a logical step to pursue in the program of expanding State park facilities.

—Staff Writer (1944)[9]

THREE AUCTIONS FOR PILOT MOUNTAIN IN 1943 AND 1944

Before the first auction, Beasley bid $10,250, leaving ten days for someone to bid more.[10] On November 12, 1943, the land of Pilot Mountain was advertised for auction and it was sold.[11] Later, another auction to sell Pilot Mountain was held on December 27, 1943, and it was announced as sold again on December 31, 1943.[12] Beasley reportedly bid $10,750 for the second auction and this was said to be the highest bid.[13] What happened with the first two auctions is unclear, but a third one was scheduled for 1944. It's highly plausible that this third auction occurred because whoever had the highest bid might not have been able to come up with the required one-third cash down payment. Of course, these three

9. Staff writer, "Under the Dome," *The News and Observer*, January 22, 1944.
10. Staff writer, "Beasley Bids $10,250 for Pilot Mountain," n.s., October 19, 1943, Mount Airy Historical Museum.
11. Staff writer, "Pilot Mountain Auction Is Set for November 13," *The Charlotte Observer*, November 12, 1943; Staff writer, "Beasley Bids $10,250 for Pilot Mountain," *Mount Airy News*, November 19, 1943.
12. Staff writer news item, "Pilot Mountain Sold on Monday," *Mt. Airy Times*, December 31, 1943.
13. Staff writer, "Pilot Mountain Sold on Monday," *Mt. Airy Times*, December 31, 1943.

auctions were held in the midst of WWII. The flyer announcing the auction described the sale:

On **SATURDAY, NOVEMBER 13th, 1943**, a tract of about 1060 acres embracing the Pilot Mountain will be offered for sale at the gate on the road leading from Pinnacle, N. C. to the top of the mountain. Sale will be made at 12:30 p. m., to the highest bidder on the following terms: 1-3 cash, 1-3 in six months, and balance in 12 months, deferred payments bearing interest.

This October 6th, 1943.

J. S. COOK and JOHN R. HOFFMAN,
Commissioners.

(PLEASE POST UP)

FIGURE 6.7. This is a portion of a flyer for the first auction of Pilot Mountain.

The buildings listed on the flyer for the second auction no longer exist (fig 6.8). Only the foundations remain visible on the property today. It is interesting to read the description of the deed survey where it states "to a rock pile" or to "that pine tree" or to the "red bud and a rock pile." Now, over eighty years later, what does the "small oak" look like as a point of reference all these years later?

The third and final sale of Pilot Mountain took place as a "re-sale" on Monday, February 7, 1944, at 12:30 p.m. The third price set for the auction was $11,287.50.[14]

Figure 6.8 is the auction flyer from the second of three auctions after Spoon died, so it wasn't determined who would own the mountain until the next year.

At Spoon's final auction in early February of 1944, William E. (Big Will) Matthews had the highest bid for the mountain, but Beasley entered an upset bid of $14,500, purchased the property, and then took possession of 1,066 acres of land. That acreage included the Little and Big Pinnacle of Pilot Mountain.

> Aside from sentimental interest, there is a commercial interest attached to the [Pilot] mountain most people know little about. In the 1,066-acre tract making up the mountain, there is a great deal of good tobacco land-fine gray soil found in Surry and Stokes counties. Although only four acres of AAA tobacco allotment goes with the mountain, much more than that acreage is potential tobacco growing land. And despite the repeated forest fires that have swept the mountain from time to time, there is valuable timber on the slopes and a vast amount of wood.
>
> —Staff Writer (1943)[15]

---

14. "Another Re-Sale of The Pilot Mountain!" *The Mount Airy News*, January 28, 1944.
15. Staff writer, "Mountain to be Sold at Auction: Pilot Landmark Part of Estate," *Daily News*, November 10, 1943.

On March 31, 1944, it was finally announced that local automobile dealer. J. W. Beasley, of the City of Pilot Mountain, had officially submitted the highest bid for the mountain.[16]

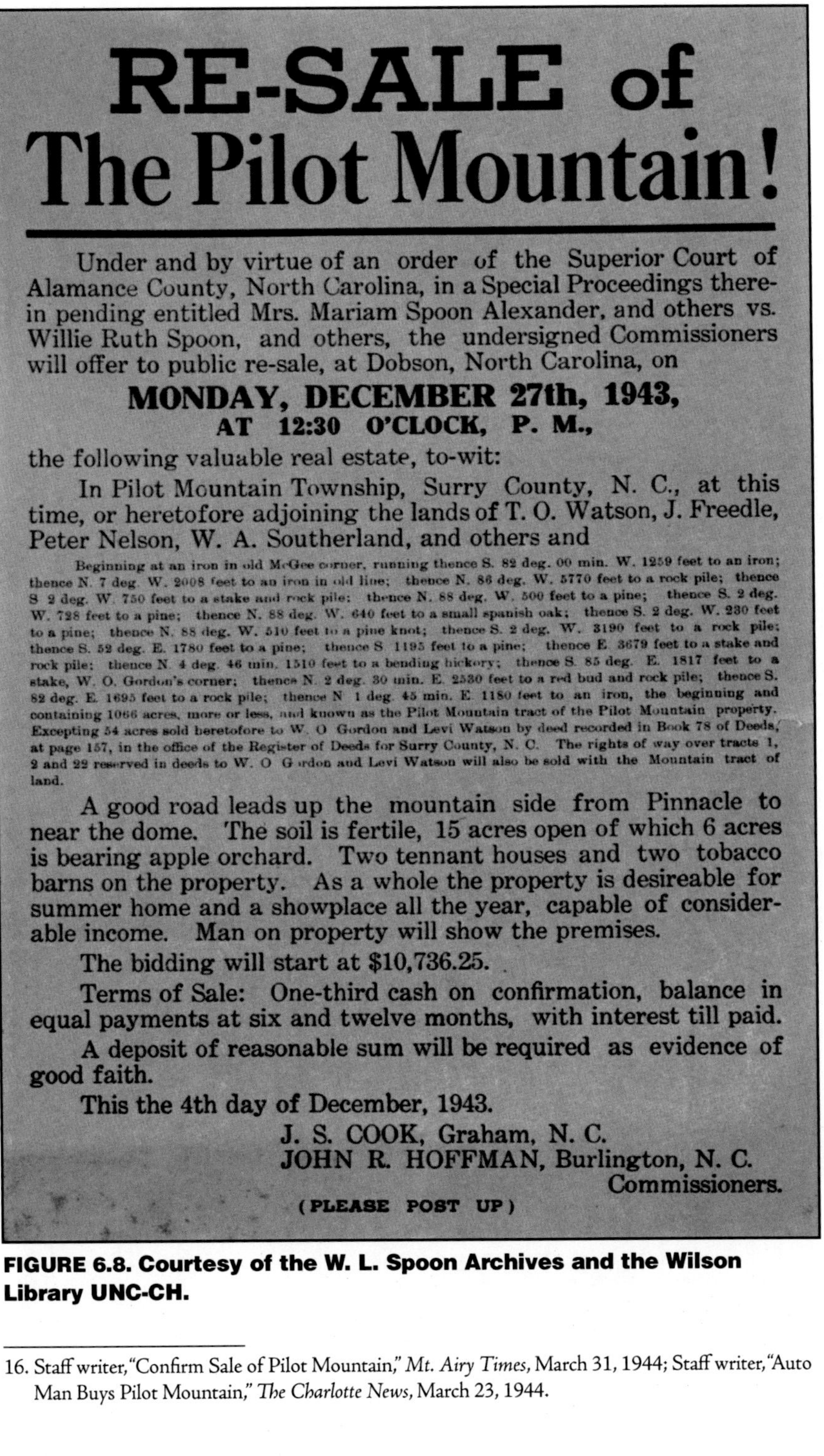

RE-SALE of The Pilot Mountain!

Under and by virtue of an order of the Superior Court of Alamance County, North Carolina, in a Special Proceedings therein pending entitled Mrs. Mariam Spoon Alexander, and others vs. Willie Ruth Spoon, and others, the undersigned Commissioners will offer to public re-sale, at Dobson, North Carolina, on

MONDAY, DECEMBER 27th, 1943,
AT 12:30 O'CLOCK, P. M.,

the following valuable real estate, to-wit:

In Pilot Mountain Township, Surry County, N. C., at this time, or heretofore adjoining the lands of T. O. Watson, J. Freedle, Peter Nelson, W. A. Southerland, and others and

Beginning at an iron in old McGee corner, running thence S. 82 deg. 00 min. W. 1259 feet to an iron; thence N. 7 deg. W. 2008 feet to an iron in old line; thence N. 86 deg. W. 5770 feet to a rock pile; thence S. 2 deg. W. 750 feet to a stake and rock pile; thence N. 88 deg. W. 500 feet to a pine; thence S. 2 deg. W. 728 feet to a pine; thence N. 88 deg. W. 640 feet to a small spanish oak; thence S. 2 deg. W. 230 feet to a pine; thence N. 88 deg. W. 510 feet to a pine knot; thence S. 2 deg. W. 3190 feet to a rock pile; thence S. 52 deg. E. 1780 feet to a pine; thence S. 1195 feet to a pine; thence E. 3679 feet to a stake and rock pile; thence N. 4 deg. 46 min. 1510 feet to a bending hickory; thence S. 85 deg. E. 1817 feet to a stake, W. O. Gordon's corner; thence N. 2 deg. 30 min. E. 2530 feet to a red bud and rock pile; thence S. 82 deg. E. 1695 feet to a rock pile; thence N. 1 deg. 45 min. E. 1180 feet to an iron, the beginning and containing 1066 acres, more or less, and known as the Pilot Mountain tract of the Pilot Mountain property. Excepting 54 acres sold heretofore to W. O. Gordon and Levi Watson by deed recorded in Book 78 of Deeds, at page 157, in the office of the Register of Deeds for Surry County, N. C. The rights of way over tracts 1, 2 and 22 reserved in deeds to W. O. Gordon and Levi Watson will also be sold with the Mountain tract of land.

A good road leads up the mountain side from Pinnacle to near the dome. The soil is fertile, 15 acres open of which 6 acres is bearing apple orchard. Two tennant houses and two tobacco barns on the property. As a whole the property is desireable for summer home and a showplace all the year, capable of considerable income. Man on property will show the premises.

The bidding will start at $10,736.25.

Terms of Sale: One-third cash on confirmation, balance in equal payments at six and twelve months, with interest till paid.

A deposit of reasonable sum will be required as evidence of good faith.

This the 4th day of December, 1943.

J. S. COOK, Graham, N. C.
JOHN R. HOFFMAN, Burlington, N. C.
Commissioners.

(PLEASE POST UP)

FIGURE 6.8. Courtesy of the W. L. Spoon Archives and the Wilson Library UNC-CH.

16. Staff writer, "Confirm Sale of Pilot Mountain," *Mt. Airy Times*, March 31, 1944; Staff writer, "Auto Man Buys Pilot Mountain," *The Charlotte News*, March 23, 1944.

> The sale of Pilot Mountain, famed round-knobbed peak in Surry County, to J. W. Beasley, automobile dealer of Pilot Mountain, has been confirmed. Beasley, with an offer of $14,500, was the highest bidder for the mountain. Confirmation of the sale was held up and the mountain went under the hammer three times. On Monday of last week the sale was confirmed in Alamance County Superior Court.
>
> Beasley said he expected to make the mountain an important resort center; but that of course his plans would be held in abeyance until after the war. 'There already is a road up the mountain and this will be improved and maintained,' he said 'I also expect to build an electric line up the mountain and provide it with lights. Some construction also will be done.
>
> —Staff Writer (1944)[17]

EVENING OF BUYING PILOT MOUNTAIN AT THE BEASLEY HOME

> My mother and I were at Granny's when Grandpa drove in the night (to tell us) he bought the mountain. He loved auctions – loved buying property, fixing it up and reselling it! Too tired to call her 'honey, I'm home'... He came in the side parlor door and I can just see him coming in.
>
> —Betty Gay Shore (2020)[18]

What followed is something Beasley's oldest grandchild, Betty Gay Shore, will never forget. Betty Gay, who was almost six when her grandpa came home with some big news, said she heard the whole exchange. This was on the Saturday night after the auction at Pilot Mountain's entrance gate on the clay and sand road. According to Betty Gay, Beasley and his wife had the following conversation:

"Where have you been, because it is after dark again, and what did you buy today?" her grandmother, Pearle Beasley, asked when he arrived home. She continued doing exactly what she had been doing before he came in the door. He always had bought something, so it wasn't an unusual question to ask, even though it was a very unusual answer.

"I bought a mountain," he responded.

"What mountain?" Pearle asked.

"Pilot Mountain," he replied.

---

17. Staff writer, "Confirm Sale of Pilot Mountain," *Mt. Airy Times*, March 31, 1944.
18. Betty Gay Shore, phone interview.

"Why in the world did you do that, John?" asked Pearle. She was concerned because he had rheumatoid arthritis and his swollen hands made it difficult for him to drive, especially up the mountain's winding road.

"The man who bid at the auction wanted to put a cable car up the mountain to the top and that would ruin it," said John. "You know I've always loved that mountain and they had that auction today. They wanted to put a cable car up to the top. That beautiful mountain! They would ruin it with cable cars and I couldn't let 'em do it."[19]

> That night he told us he had bought Pilot Mountain Granny cried; she was always crying about things Grandpa did. My mother hugged him when he told us 'what mountain' he bought. We later learned of the upset bid – he loved that mountain!
>
> —Betty Gay Shore (2020)[20]

Betty Gay also remembered her grandfather's will specified that the mountain could not be sold for any commercial venture.

> I was told that in the will he stated that it be kept in its 'natural state.' It was not to be sold for commercialism – (how true I did not see the will?) – but would have believed it and still believe it! That was my Grandpa – a smart, simple, loving man.
>
> —Betty Gay Shore (2020)[21]

BEASLEY UPGRADES TO PILOT MOUNTAIN

The Beasley family developed Pilot Mountain for several years after the purchase by paving Spoon's earlier sand and clay road. It was announced in April of 1946 that the four-mile road bed to the top of Pilot Mountain had been prepared for a thirty-foot-wide black-top highway. It was completed in just sixty days after the newspaper announcement.[22]

---

19. Shore, phone interview.
20. Shore, phone interview.
21. Shore, phone interview.
22. Staff writer, "Improvements on Pilot Mountain," n.s., April 12, 1946, Mount Airy Historical Museum.

> Grandpa cleared up a spring on the left side of the road going up the mountain, as cars were not built then to handle the steep grade of the road to the parking lot. More cars than not, were seen with stream coming out of the radiators on the side of the road.
>
> He cemented the area around the spring and left 'one' dipper by it every day – just one – it was always gone by the end of each day – I remember. Mama would not allow me to drink from the dipper – but I could cup my dirty hands and drink from the cool, so cool, spring water that trinkled from the spring – God's nectar!
>
> —Betty Gay Shore (2020)[23]

The building of the pool turned out to be a little tougher than was originally imagined:

FIGURE 6.9. Pool on the side of Pilot Mountain circa 1947 to 1968. Now, it would take an archeological dig to find it. Courtesy of the Frank Jones Photography collection at the NC Room of the Forsyth Public Library in Winston-Salem, NC.

23. Shore, phone interview.

> The pool, which is being constructed on a slope of the mountain, is no little engineering feat in itself. When it is ready for use, it will be one of the highest pools in this part of the State.
>
> The new swimming center is being built by the B & L Construction Company of Winston-Salem with all of the concrete being hauled from the Twin City. Work was started early this summer with a goal of having it ready for use about 30 days ago. But building a swimming pool on the side of the mountain is something that cannot be done on an exact schedule. J. W. Beasley, Pilot Mountain automobile dealer and owner of the mountain, said that the pool will be ready for use with the beginning of the season next Summer, though... The steps to the peak of the knob have been improved by Beasley and a flagpole stands like a sore thumb atop the rock and tree-covered knob itself.
>
> But the 50 by 75-foot swimming pool, its bathhouse, lake and dam are the culmination of Mr. Beasley's planning for Pilot Mountain.
>
> —State Writer (1949)[24]

A dam was also constructed to form a mineral-spring-fed pond for the pool. This little pond was stocked with fish, and nearby, there was a six-acre apple

FIGURE 6.10. One of the many old roads J. W. Beasley built in the 1940s, now used as a trail in the campground area.

24. Staff writer, "Another Modern Touch Being Added to Mountain Landmark, New Swimming Pool Will Be Ready for Use by Next Summer," *Winston-Salem Journal and Sentinel*, September 10, 1949.

orchard as described by the Beasley grandchildren in the first chapter. The apple orchard was also mentioned in the W. L. Spoon auction flyer. It is unknown who planted the apple orchard, but the Native Americans in this area often had extensive orchards. At first, the mineral spring water was intended to flow down to the pool through a specific natural filtering process. But it later was determined chemicals had to be used to keep it clean from so much use.[25] The pool became a popular gathering place for the city of Pilot Mountain and the surrounding area.

Despite Pearle Beasley's original teary response, Betty Gay reported in a letter from September 2021 that Pearle Beasley worked on Pilot Mountain often, confirming what her cousins had said on the walk through the park in chapter 1:

> Grandma built scores of rock walls; Her passion was rock walls.
>
> —Betty Gay Shore (2020)[26]

In 1947, the Western Union and Telegraph Company bought an acre on the top of Pilot Mountain to use as a "radio relay station." This is also noted in the *Winston-Salem Journal and Sentinel* report. That acre is said to be near the parking area.

SCENIC WAYS TO PILOT MOUNTAIN ON THE BACK ROADS

After reviewing some of the book chapters about her family for accuracy, Betty Gay Shore mentioned,

> Still have seen nothing telling anyone how to get to the mountain unless exiting #52. There are several scenic ways – from King; we went old #52 to Pinnacle, exited by going through Pinnacle proper at the railroad crossing to another "old US Route 52," crossed the tracks then turned left toward the mountain on a winding beautiful old road, that spanned curves and on left one lone lovely meadow I painted with Indian paint brush. So cool in the summer, that Pilot Mountain football team practiced there. Then an incline toward the gate that had the old souvenir shop on the right [now on the new road it is on the left but still features Mrs. Pearle Beasley's rock walls].
>
> —Betty Gay Shore (2020)[27]

25. Glace, Glace, and Sperry, interview.
26. Shore, phone interview.
27. Shore, phone interview.

FIGURE 6.11. Pilot Mountain owner, J. W. Beasley, and his daughter, Carole B. Sperry. The ladder up to the top of the Big Pinnacle is located to the left of J. W.'s hand. Behind his hat is where the notch in the Big Pinnacle is located, but it is obscured by thick trees. This appears to be the Little Pinnacle overlook before the wooden fencing. Courtesy of Patricia Sperry.

There is a spot on the road near the top of the mountain, just before the hairpin turn into the parking lot, where J. W. Beasley liked to stop and just look at the Big Pinnacle and the Sauratown Mountain beyond it. Beasley always kept the trees there cut down, so even before you parked your car, you had a magnificent view. That road was much closer to the edge of the cliff than the current Pilot Mountain State Park road, but it's still a great view.

> You know when I was growing up, there was a sharp turn to the right at the top of the road just before you get to the picnic area at the top. Grandpa wanted everybody to stop right on the road. He would keep those trees cut down because from there you just wanted to stop. You could look directly at the Big Pinnacle and it was beautiful. Oh, it was just beautiful. Anytime we ever went up there, we always just gasped when you go on up the road and all of the sudden you are going to be turning right and you look at the Pinnacle in front of you... I always gasped at seeing the Big Pinnacle on the left – spectacular – God's creation!
>
> —Betty Gay Shore (2020)[28]

28. Shore, phone interview.

THOSE "RICKETY STAIRS" UP THE BIG PINNACLE

"Go at your own risk," warned a sign at the base of the stairway on Pilot Mountain for many years.[29]

As the years went by, the ladders were replaced by stairs by Spoon and then those stairs were upgraded or replaced by Beasley. The ladders and stairs were often referred to as "rickety" the older they got. It didn't really matter if they were ladders or wooden stairs, both attracted lightning, as if they were lightning rods. Betty Gay recalled those stairs:

> My dad put me on his shoulders and climbed the stairs to the Big Pinnacle. I don't know how old I was but my poor mama at the bottom of the stairs was in tears, as she was so frightened. I first beheld the sight before my eyes – nothing short of magnificent I would guess, nor had it changed into my teen aged years... I understood why they removed the stairs but cry even now, at 82 years old, that more could not have witnessed the beauty on a clear day.
>
> —Betty Gay Shore (2020)

It was reported that after Beasley died, his widow, Pearle, was offered bids to buy Pilot Mountain by motel chains and oil companies, but she held fast to her husband's wishes and did not sell. The Beasley family set a straight course to keep the mountain in its natural state and they did not waiver, even after Beasley passed. There were tempting offers, but it was not what the family wanted for the mountain. Everyone who visits the mountain should know how steady and sure that family was for Pilot Mountain to remain primitive and not cave to commercial interests.

LANDING A PLANE ON TOP OF PILOT MOUNTAIN

> Grandpa is the one who built the place for the plane to land. Grandpa had bulldozers and decided that he wanted the people who had little planes to be able to land there and go hiking. He just wanted to spend some money and that's what he wanted to do. He just wanted this project. Anyway, he decided that he was going to pay for an airfield for a plane to land... I remember the dust, most of all, not red clay dust, light beige colored dust flying everywhere. Grandpa was the 'quiet man' but his

29. William L. Spoon Papers, images folder.

FIGURES 6.12 and 6.13. The airfield on Pilot Mountain in 1946 (above) and almost the same view in 2021 (below). Image above courtesy of the Frank Jones Collection at the Forsyth Public Library.

excitement that day was palpable!! So, grandma thought it was crazy. Everybody in the whole family thought that was crazy but on the day that the plane landed we had to be there. Everybody that he could round up, from the whole family, had to go to the top of Pilot Mountain to see that first plane land. It was a must! Not an option! This gentle quiet man was so excited. So, we all went to the mountain and that plane landed on that short landing strip. I don't know whether another plane ever landed there or not? I can show you right now. I know exactly where it was but it was just a short landing strip and we thought the planes could go into the trees.

—Betty Gay Shore (2020)[30]

In November 1946, many of the upgrades Beasley had made to Pilot Mountain were announced in the Henderson Daily Dispatch:

Mr. Beasley lives in Pilot Mountain and owns Pilot Mountain itself. That is, he owns the 1,066-acre lump of earth and stone which rears 2,700 feet above sea level in the Piedmont plateau north of Winston-Salem and

FIGURE 6.14. This is a building in Tulum in Mexico that is known as the Temple of the Descending God. The roof is purposefully tilted at 23.5 degrees. This somewhat variable tilt is the basis of the Mayan Calendar and the great year that is roughly 26,000 years long. The tilt of the earth built into the airfield puts it in the natural or man-made theme found throughout these two books. It is always mentioned that the airfield was right around 25 degrees in newspaper articles, which, of course, made it difficult to land at the makeshift airport.

30. Shore, phone interview.

> which may be seen on a clear day from as far away as 75 miles. And right now, he is rushing to completion for an air park atop its narrow crest.
>
> In a few weeks or months, perhaps those of you who have driven or flown by Pilot Mountain or have seen it picturized on almost everything from coffee cans to insurance stationery may be able to take your family plane and fly clear up to the broad part of its back for a close-up inspection and Sunday picnic.
>
> Having spent upward of $30,000 on it already, Mr. Beasley plans to pitch a string of cottages and pave the winding two-mile road from the valley up to his cloud-level park. Then he will hang out a welcome mat to cross-country air travelers and to the motoring sight-seers.
>
> Although a bit steep and still soft, the lofty landing strip has been in use already, he says, a few light planes from neighboring fields having lighted there just for the heck of it. The one runway is inclined about 25 degrees, is about 1,500 feet long, and extends from a grove of trees on one end to the top of a dizzy cliff at the other.
>
> —Ed Lewis (1946)[31]

This offhand comment that "the runway is inclined about 25 degrees" gives one pause, since that could reflect the tilt of the earth that oscillates between 22.1 and 24.5 degrees. This is another hint at the deep mysteries of the mountain's man-made versus natural phenomenon. Talking about the area where Beasley liked to stop and look at the Pinnacle just before the airfield area, Betty Gay recalled:

> We had a reunion down there but it was immediately to your right (where the picnic area is now), that was where he had that landing strip. . . Let me tell you something else. He was delighted there was a spring along that road. I can't tell you where it was now but he had it cemented in when he had the road paved. Then everybody shared the same Dipper. There was one dipper there, he had to continually get dippers because people would steal them and that was along the road on the left side going up the mountain. I know he was delighted to have that spring because he said so many people got so hot when they would get on that road. You'd see a lot of cars pulled off to the side and they would just be steaming going up that mountain.
>
> —Betty Gay Shore (2020)[32]

31. Ed Lewis, "In the Air," *Henderson Daily Dispatch*, November 8, 1946.
32. Shore, phone interview.

IS THE DEVIL'S DEN CAVE INCIDENT JUST MOUNTAIN LORE OR FACT?

A resident of Pinnacle, Gary Tilton, lives at the base of Pilot Mountain between the old Gillam Hotel and Pilot Mountain State Park. It is nice that we can now talk to people about their stories rather than only rely on written documents from hundreds of years ago, sometimes from unclear origins. While taking images on the south side of the mountain, I had the privilege to speak with Tilton. Tilton was retired by then but was an engineer in the area working on computers and radio equipment. Tilton helped install the radio towers on the Sauratown Mountains and the vibrations of that installment sent the timber rattlesnakes into the farm fields below. For a time, Tilton was not welcome at the midmorning coffee klatch among the farmers in King, North Carolina. We spoke in his driveway on a beautiful spring day, leaning against a fence in his yard for hours. The thermals off the stone rock were causing havoc with the focus of the camera in front of his home, making for a constant blurred image, so a break was welcome. Tilton thought it was about time someone wrote a book on Pilot Mountain, which he clearly had a fondness for.

"Have you heard the story about the three boys who got in trouble on the mountain?" Tilton queried, leaning against his garden fence in his driveway, in a discussion about some of the unusual aspects of Pilot Mountain.[33]

> They found a cave they could fit in and saw inside there was a vertical shaft going down into the mountain. They went and bought some rope so one of them could shimmy down into the mountain to see how far down the vertical shaft went into the mountain.
>
> After getting the rope one of them tied it around their waist and the other two held the other end tightly. The boy had gone down a good little way when the rope went slack like it was cut. They brought it up and the rope, indeed, was cut, most likely by a sharp rock on the way down. They heard yelling at first so one of the two other boys attempted to go down with what was left of the rope but it wasn't long enough to get to the bottom.
>
> The sheriff was called but they couldn't hear any signs of life later when he arrived. So they used dynamite to seal the cave entrance so nobody else would wander into it since it was a clear danger to the public.
>
> —Gary Tilton (2020)[34]

33. Gary Tilton, interview with author, 2020.
34. Gary Tilton interview in person 2020.

Tilton had no specific date for the cave story. Members of the Surry County Sheriff's office and the Pilot Mountain Search & Rescue also hadn't heard of this incident. However, the Devil's Den no longer has access to a cool large room inside the tight opening. One can no longer crawl into it to the ledge within, which appears in reports over the past two centuries. This might just be another story in the lore of Pilot Mountain, but consider that many at first didn't believe there was one airfield on the top of the mountain, let alone two versions during different decades. Further research is needed for this interesting but tragic turn of events. The fact that the Devil's Den is no longer a cave you can squeeze into gives one pause about this story.

Tilton also mentioned that when he built an out building as an office, he noticed a curious thing that happens with voices on the mountain. He could clearly hear people talking from over a mile away on the south side from the concave on the Pilot Knob Trail. It was like those people were talking in their normal voice right there in his office. When his grandchildren visit, they like to test this out with the long-range mountain voice and his office. There is a room in the Morehead Planetarium complex at the University of North Carolina at Chapel Hill that, while on a smaller scale, has the same effect. You can whisper in a crowded noisy room and hear it in the other corner way across the room, like someone is right next to you.

PRICE WAS PER CAR TO GO UP PILOT MOUNTAIN

When grandpa owned the mountain, the price was as many as could fit in the car – one price only per car. Just to see, on one July 4th, when I brought thirteen nursing school classmates' home with me (as most were from the northern states and could not go home for the holidays). We crammed into mama's old green/white boxy chevy – thirteen of us – but we were much smaller in stature then for $1.00 per car!!!

Aunt Carole [Sperry] collected the money after grandpa died, she and Bert Coleman. There was money to be made in cash. Bert was the caretaker of the mountain. Cigarette hanging out of his mouth – small, angry and yes craggy – best friend grandpa ever had. Mr. Coleman and I dug up red buds from a creek to start at my house in King- Good people! Bert kept the mountain going when Grandpa died. Aunt Carole loved the mountain but Aunt Fran and my mama would take a turn at the gate when Aunt Carole couldn't (sick or needed time off). Aunt Carole lived with Granny for a long time when she came home from California.

—Betty Gay Shore (2020)

FIGURE 6.15. Painting of Bert Coleman by Joseph Wallace King. Courtesy of Bert's daughter-in-law, Winifred Coleman.

OUR MISS PEARLE OF PILOT MOUNTAIN

Pearle Beasley ran the mountain for a decade after her beloved husband passed away in 1958. According to her daughter, Carole B. Sperry, she led a very interesting life of ninety years:

> Pearle Belle, wife of John William was a nature lover, like her husband, she lived more on the outside than in. Also a collector of antiques, which the home was noted for, she never quit collecting as long as she lived. At one time, she went to Baltimore, Maryland and brought back a truckload of antiques, and brought them home to the dismay of her husband. She was born with talent... the ear for music... She could play every instrument that was handed to her; also a poet, she would take trips every year, as that was one of her loves, to travel and see the USA. She would come back from her travels and write poems of the trip and experiences along the way.
>
> —Carole B. Sperry (1984)[35]

35. Carole B. Sperry, "The John William Beasley Family," in *The Heritage of Surry County North Carolina*, vol. 1, ed. Hester Bartlett Jackson (Surry County Genealogical Society, North Carolina: 1983), 44–46.

FIGURE 6.16. One dollar per car was collected at the Gate House, often by Carole B. Sperry, in the summer months. Courtesy of Patricia Sperry.

During the depression she was appointed by the W.P.A. to work on projects in Pilot Mountain to give the destitute work, these included the Pilot Mountain cemetery and the Pilot Mtn. High School. The cemetery was made into a beautiful place by her and the workers. Her handiwork is still visible today around the home, cemetery, school and First Baptist Church (of which the family was a member). She received from *Better Homes & Gardens* Honorable Mention for doing outstanding work in Civic Beautification. The plaque which she received read, "The project for which the award is made has required long and painstaking effort and a rare spirit of cooperation. The achievement has stimulated nationwide interest in community betterment as well as in individual home gardens for which the community forms a necessary and fitting background. This award is made at the conclusion of the More Beautiful America Contest, conducted during the period beginning Sept. 1, 1932, and ending Oct. 1, 1934." This contest was sponsored through the Garden Club of Pilot Mnt., of which Mrs. Beasley was a member.

In 1946 Oscar W. Smith, Mayor of Pilot Mtn, proclaimed Pilot Mtn. a Bird Sanctuary through the urging of Sir Guy Thomas, then a resident of the city. Miss Pearle, as many of her friends call her, decided to write a column in the local newspaper. *The Pilot News* on birds and it was called,

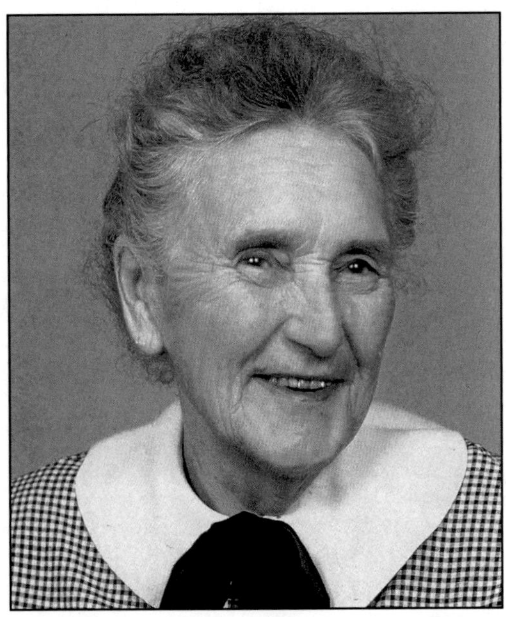

FIGURE 6.17. Pearle Beasley (a.k.a. Miss Pearle) refused many attempts to buy Pilot Mountain for commercial purposes, fulfilling her husband's wish to keep it natural.

> 'Bird Life.' Her first column was 'Town of Pilot Mountain Made Bird Sanctuary." Her articles were very interesting and educational.
>
> —Carole B. Sperry (1984)[36]

Carole B. Sperry was very respected in the community, and traditional musician Doc Watson sang at her gravesite when she passed. Sperry often would paint Pilot Mountain on postcards, and newsman Charles Kauralt received some of them. In a thank you note back to her daughter after she had passed, he wrote:

> I am very glad to have the examples of your mother's work. I'll never think of Pilot Mountain without thinking of Carole Sperry. Thank you for your thoughtfulness.
>
> —Charles Kuralt (1993)[37]

36. Carole B. Sperry, "The John William Beasley Family," in *The Heritage of Surry County North Carolina*, vol. 1, ed. Hester Bartlett Jackson (Surry County Genealogical Society, North Carolina: 1983), 44–46.
37. Charles Kuralt, private letter to Patricia Sperry.

PILOT MOUNTAIN RECAPTURES SIGNAL MOUNTAIN NAME IN 1947

A Pilot Mountain scrapbook featured an undated newspaper article that was written by Bill East and published by the *Winston-Salem Journal and Sentinel*:

> Decades ago, according to legend, Indians once flashed signals from the stone-capped knob of Pilot Mountain. From that vantage point, the orders of the early Americans could be seen plainly and relayed on to the next group of tribesmen... The system worked like a charm. It worked so well, in fact, that the Northwest North Carolina peak became known as the 'pilot'—or guide—of those who came near it... The Western Union Telegraph Company today confirmed reports that the mountain has been selected for a new role. The new part that Pilot will play in the development of North Carolina and the South will be akin to the role it served with the Indians [Native Americans]. The company said that Pilot Mountain had been selected as the tower site for a radio relay station... In view of the rapid advances being made at this time in microwave technique, we are unable to give at this time factual information concerning the type of equipment, allocated frequencies, transmitted power levels and tower constructions features.
>
> —Bill East (no date)[38]

The Southern Mapping and Engineering Company was involved in the exact location of the relay and did the work on May 27, 1947. The right-of-way up the mountain and a little less than an acre was sold to Western Union Telegraph Company on September 22, 1947, for one-hundred dollars. From that point forward, Western Union could repair the Pilot Mountain road themselves to ensure it was "open and passable."[39] Pearle Beasley wrote a letter to the Western Union Telegraph Company in 1963 about repurchasing the track of land they had bought in 1947.

The response came from Western Union Telegraph's general manager of real estate on April 8:

> Upon receipt of your letter we checked with our engineers to determine whether this site was still to be used for our Microwave network and we were informed it is to be used and could not be abandoned,

38. Bill East, "Pilot Mountain: Chosen as Link in System of Rapid Radio Communication," *Winston-Salem Journal* and *Sentinel Newspaper*, n.d.
39. Western Union, letter, Ms. Pearle Beasley Private Archives. Written from a response by M. G. Halligan, Atlanta, GA, area superintendent of Western Union.

> In these circumstances we could not sell this property to you. We have, however, made a copy of the deed and it is enclosed for your perusal.
>
> —R. V. Wagoner (1963)[40]

This less than an acre of land owned by the Western Union Telegraph Company on top of Pilot Mountain was mentioned in the title search when the Pilot Mountain State Park paperwork was signed.[41] The acre of land and right-of-way owned by the Western Union Telegraph was made clear in the "option to purchase" in the last few paragraphs of the agreement on August 25, 1967, where ten dollars of earnest money was exchanged from the Park Committee to Pearle Beasley.

ZACH REYNOLDS STUNT PLANE WORK

More lore on the mountain included a red stunt plane that used to fly over Pilot Mountain practicing dives, stalls, twirls and buzzing under the Pinnacle overpass

FIGURE 6.18. This is the Pinnacle Bridge on US Route 52 that both Zach Reynolds, and years later, Jimmy Dean flew under in their flatwing airplanes. Traffic was held back when Jimmy Dean flew under the bridge for an independent film, but Zach Reynolds just flew under in the midst of traffic. Sadly, a plane he was in crashed in the woods near this bridge and all were killed.

40. Western Union.
41. Pilot Mountain State Park Deed, Private Beasley Archives and Surry County Deed office, Dobson, North Carolina.

on US Route 52 often. This happened over the course of many years. Like the airfield on top of Pilot Mountain and the Devil's Den stories, it may be hard to believe at first. Mostly, old timers would mention it in passing with a few more details in each telling. It was often told at a stop for gas in Pinnacle off of US Route 52 by an old timer having a smoke. After tending chores on the farm, they were ready to talk to someone eye to eye. As more people were interviewed for the book, they all started to tell the same stories about a daredevil stunt flyer in the red plane. One evening, another granddaughter of Pearle Beasley, Patricia Sperry, mentioned the stunt flying over Pilot Mountain, but this time it was by name in some detail. She recalled:

> Zach [Reynolds] always wanted to stay on grandma's good side. Zach would call my mother at her school, Pinnacle Elementary, and from the Winston-Salem airport he told her to have her 5th grade class take recess in about twenty minutes. Then while they were outside, he would fly over Pilot Mountain doing his stunts to practice for a state fair or something.[42]

These same type of stunt plane maneuvers by Zach Reynolds, grandson of R. J. Reynolds, were filmed by Joseph Wallace King for his movie, *Somebody Moved My Mountain*. The stunts were performed by stunt flyer Jimmy Dean, who owned the exact same kind of plane as Reynolds. While Pinnacle Elementary would have a great view for that, King placed his camera on the Little Pinnacle to film the stunts for an even closer view, which is now available on YouTube.[43] This seems to be clearly a fictional tribute to the real-life Reynolds.

What was the reason for so many practice visits for stunt flying by Reynolds around Pilot Mountain?

> Pilots perform their sequences in the 'aerobatic box,' an imaginary cube of airspace whose sides measure 1,000 metres (3,300 feet), with a minimum lower safety limit below which pilots may not fly. Penalties are imposed for flying outside or below these limits. A pilot is expected to perform each sequence correctly, accurately, and precisely, and further penalties are assessed when any kind of error is made.
>
> FAI [English version is: International Aeronautical Federation] competitions usually aim to test competing pilots with a number of different sequences, which may include a known compulsory (announced and practiced in advance), one or more unknown compulsories (flown without practice), plus one or more freestyle sequences, which each pilot may design individually within certain constraints.

42. Glace, Glace, and Sperry, interview.
43. Joseph Wallace King, director, *Somebody Moved My Mountain*, 1975.

> For the purposes of competition, aerobatic maneuvers have been codified in the internationally accepted FAI Aerobatic Catalogue, which gives a point value to each maneuver. These basic maneuvers may be flown on their own (e.g., a single vertical roll) or may be grouped into complex figures containing several maneuvers (e.g., a stall turn may begin with a vertical half-roll on the climbing line, followed by the stall-turn rotation at the top, followed by another half-roll on the descending line.
>
> —Annette J. Carson (2017)[44]

EARLY ROCK AND ROLL ERA DANCES ON THURSDAY NIGHT

"Everyone who was anyone was there!" Betty Gay Shore recalled. "We would have walked rather than miss a pool night at the covered dance area. A way to meet boys—or for the boys—a way to meet girls from the surrounding towns."

Anyone who had been to the Pilot Mountain dance pavilion remembers those Thursday nights fondly, including the band members, the Bert Coleman grandkids, and all Beasley cousins in this book. Billy Long and the King Bees, whose leader is from just down the road in King, NC, were the house band almost every Thursday night. People came from miles around to dance till midnight, and many strong friendships were formed with memories to spare.

"Not sure they ever played anything other than current rock and roll hits," is the way Kenny Glace Jr. remembered it. Part of what created the edge for the King Bees, who often dressed in yellow and white suits, was a DJ in Winston-Salem who gave them recordings of the week's top ten 45s. By Thursday, they knew the

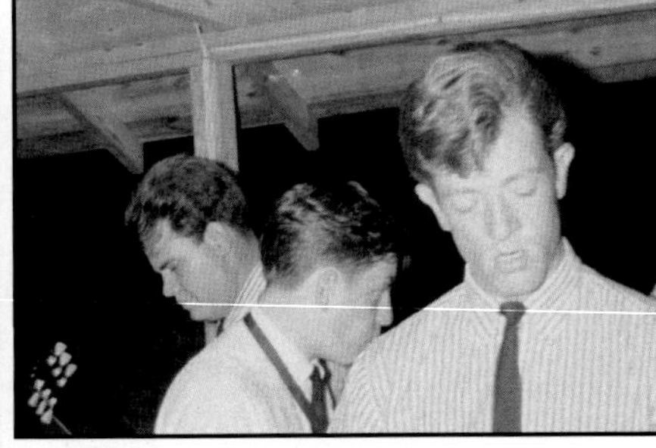

FIGURES 6.19 and 6.20. King Bees playing on Pilot Mountain in the Dance Pavilion. Images courtesy of Patricia Sperry.

44. Annette J. Carson, "aerobatics," in *Encyclopedia Britannica*, December 21, 2017, https://www.britannica.com/sports/aerobatics.

top hits for that week and played them. Kenny Glace was a lifeguard at the pool during the day and after the pool shut down in the late afternoon, the dance pavilion opened each Thursday night at eight. "Of course, Friday morning meant you had to clean up the parking lot with all the soda cans and such," said Kenny.[45]

R. M. Collins and his wife managed the pool and he came up with the idea of a Thursday night pavilion dance. Betty Gay Shore reminisced:

> I closed my eyes to recall the covered dance pavilion since you don't have a picture of it. It was just a cover with benches around all four sides. At one end, the King Bees performed, there was not a wall – all just covered. To walk into the pavilion, on a pole to the left of the non-covered pavilion, was a juke box. So, on Thursdays, always on Thursday nights, when there was no live entertainment we used the juke box. When we couldn't convince our parents we needed a car to go to Pilot – we could beg, borrow or steal to get someone to take us. It was our event of the summer. I met more Mt. Airy boys! UNC students – get more summer dates – to think now how dangerous it was – but really wasn't then. Got more rides home with really nice good boys (young men).
>
> The band would start setting up around seven and then take the stage for the dancing which would go till midnight. Usually, two off duty Pilot Mountain police officers were hired for security and selling tickets.
>
> —Betty Gay Shore (2020)[46]

"We worked for half the gate, but a lot of the kids came in the trunk of cars so there would be way more kids than the gate receipts," the red-headed King Bees keyboard player and vocalist Dale Riddle remembered.[47] "All the other band members have passed on now. The parking lot was tiered and the dance pavilion wasn't big enough for everyone so they just crowded around." While the band was known to be from King, only Billy Long was from there. The rest of the band would assemble from all directions at the Pilot Mountain pool area.

Like many bands, Billy and the King Bees went through some different band members over the years. Long, the band leader for the entire run, could play fifteen instruments. Jim Lowry was the impressive guitar player who would later become the band leader for country singer Donna Fargo. Lowry hailed from Mount Airy but spent twenty years in Nashville after being in the King Bees.

Billy and the King Bees had some extraordinary breaks that most bands only dream about. Long always wanted to work at his dad's jewelry store at 231 South Main Street in King, NC. Long Jewelry in King has been run by four generations

45. Glace, Glace, and Sperry, interview.
46. Shore, phone interview.
47. Dale Riddle, phone interview with author, 2021.

of the Long family now, so it really is a family tradition. Music was a hobby for Long, as he played part-time for some extra money.[48] It was nothing more and nothing less for Long in terms of his musical hobby. That was fine, except not only did they impress crowds with their musicianship, they also impressed other musicians, DJs, and record company executives.

At some point, soul singer Otis Redding took a strong interest in the band and got them an audition for his personnel label, Stax/Volt Records. Redding would come to their Atlanta club gigs and encouraged them nearly every time they played there. When a box set for Stax Records came out, it included the single "Bongo" and the B side "Susie Q," played by Billy and the King Bees. Nobody at the label remembered the band at all. The executive director at Stax Museum of American Soul Music in Memphis explained via an email.

> We sure don't have anything about Billy and the King Bees here at the museum – we typically can scrounge up some good info on the random, one-off bands post 1968, but during the early days, they didn't keep very good records of sessions and such. Sadly, the musician's union lost all of the session records pre-1967, so there's no way to even go through those.
>
> —Jeff Kollath (2021)[49]

> I thought it was interesting that Stax did not really know who we were, but I am not surprised. We drove from Columbus, Georgia to Memphis and we were there only one day trying to put something together. We were totally unprepared and had no material ready. *Bongo* was just improvised on the spot. I was surprised it was ever released because we were just a group passing through that Otis Redding had recommended to the label. Billy [Long] had no interest in becoming a full-time musician, so no effort was ever made to get a recording deal or contract. We did all meet with the manager of some of the Stax and Volt artists, but Billy let him know he was not interested in signing a contract.
>
> —Dale Riddle (2021)[50]

The producer of that Stax recording in Memphis was also helping put together a tour of America for a new English group when the audition happened in 1963. An offer was discussed that day in Memphis. Would the King Bees consider opening for some or all of the English band's tour dates? Long turned them down flat, just as he had the recording contract. Both the recording and concert

48. *Long Jewelers of King*, http://www.long-jewelers.com/.
49. Jeff Kollath, email correspondence with author, 2021.
50. Dale Riddle interview 2021.

contracts were still given to Long to reconsider, but he never filled them out. They just sat on his desk in King, North Carolina, unsigned.

The cross-country tour they turned down would have started about six hours from their home base in the Pilot Mountain area, at the Washington Coliseum on February 11, 1964. One wonders: had they said yes, would they have gotten a shot at the Ed Sullivan Show in New York? Yes, as you might have already guessed, the King Bees turned down a recording contract arranged by Otis Redding himself for his personnel label called Stax/Volt, and then they turned down the opening act spot for the Beatles' 1964 *Tour of America*. The term "Beatlemania" was coined during that tour. From a hotel room in Atlanta, the King Bees watched perhaps the most famous of the Ed Sullivan Shows, which introduced the Beatles. How different all their lives might have been had he said yes. However, to this day, Dale Riddle thinks they made the right choice.

THE FIGHT THAT ENDED THE DANCES IN 1966

Sadly, and tragically for all concerned, a fight broke out during one of the Pilot Mountain dances while a crowd danced to the King Bees' music. Pearle Beasley always hired off-duty policemen as dance security. In short order, the two young men in the fight were taken to the Pilot Mountain jail by the two off-duty police officers. As they were being put into a cell by another officer stationed at the jail, or even after they had been placed in the cell, a deadly fight broke out. Stories, even forty years later, continued to be published in area papers about the odd incident.[51] The police officer was killed with his own weapon, but then the suspect was released on bail. It didn't make sense then nor now. Usually, someone who does that is considered a danger to all of society and wouldn't be released on bail. Later, that suspect was killed in Virginia, so there wasn't a trail to confirm any of the details for either side.

"This certainly had an impact on the Thursday night dance on the mountain and the event was soon canceled following the murder," Dale Riddle wrote to me.[52] Riddle knew the man and went to his funeral. Months after this incident, the fundraising to purchase Pilot Mountain to become a state park began in earnest.

> When Beasley owned Pilot, many offers were made by firms to locate businesses on the mountain – offers that for a certainty would have guaranteed a yearly income for life. But at no time did John Beasley intend to commercialize his mountain. He wanted it kept in its natural state.

51. Scott Sexton, "Jail Cell Mystery: Sister Determined to Change View of 1966 Tragedy," *Winston-Salem Journal*, November 12, 2012, updated April 9, 2019.
52. Dale Riddle, email correspondence with author, 2021.

> After [J. W.] Beasley's death, his wishes were carried out to the fullest. Not one inch of the land was sold, although great amounts of money were offered. Motel chains wanted to build... oil companies wanted to build... and offered a staggering amount of money for a minute piece of land. A rock quarry could have been established there, and of course; tons of rock would have been blasted away.
>
> —Linda Hopkins (1969)[53]

Here we have two examples of upright people who valued life and honor above money. Billy Long managing the King Bees and the whole Beasley family honoring grandfather, J. W. Beasley in death, making sure the entire land went to forming a state park and would never again be sold for commercial purposes:

FIGURE 6.21. J. W. Beasley and Bert Coleman at Christmas time. Courtesy of Patricia Sperry.

53. Linda Hopkins, "Men and a Mountain: The Men," Enterprise Publication, ca. 1969, Pilot Mountain Library and Forsyth Public Library, Winston-Salem, North Carolina.

The family, in selling the mountain to the state, felt that would be an extension of the late owner's wishes. And the state of North Carolina has promised that the mountain will always remain exactly as it is today. If the mountain is mutilated in any way, it will automatically revert to the heirs. In this way the family feels it has contributed something worthwhile and lasting in their lifetime for humanity... Adjectives have been repeated and worn out when people talk about this mountain. And for me, it is easy to see why John Beasley truly loved this old mountain. Here is relaxation that only comes from the peace of nature. You have none of the entertainment racket. There is nothing to lure you. The mountain, quite obviously, is the only lure that is needed.

—Linda Hopkins (1969)[54]

FIGURE 6.22. Ben Coleman working a tobacco field with a helper right where the Pilot Mountain Visitor Center is now located. Courtesy of Mrs. Winifred Coleman.

54. Hopkins, "Men and a Mountain."

FIGURE 7.1. Stone face on a Pilot Mountain State Park trail.

7

Long Road to Become Pilot Mountain State Park in 1967

FIGURE 7.2. When the mountain was privately owned, this turn was a unique view of the Big Pinnacle and the Sauratown Mountains. The road has been changed since it became Pilot Mountain State Park and is much further back from the edge of a cliff than it used to be. The Beasley family kept this special view, but now trees block it.

> Prior to that time, the mountain was a private commercial tourist attraction. Pilot Mountain Preservation and Park Committee proposed the establishment of Pilot Mountain as a State Park unit for conservation, natural resources protection, and the surrounding area from future commercial development. Working with the conservation-minded owner of the property, Mrs. J. W. Beasley, the group secured options on the land and raised matching funds to purchase the land with federal grants. The committee acquired more than 1,000 acres of land along the Yadkin River that was added to the park in 1970. Pilot Mountain stands as a monument of a concerned citizen dedicated to preserving the exceptional natural resources of North Carolina.
>
> —Kevin G. Stewart and Mary-Russell Roberson (2008)[1]

FUNDRAISING EFFORT TO BECOME PILOT MOUNTAIN STATE PARK

No one can know all of what was in Pearle Beasley's mind when she decided to sell the mountain to become Pilot Mountain State Park. It was likely because of her late husband's wish for it not to be sold for commercial purposes, but why at that time? Taxes, liability because of fights at the dance, and the burden of day-to-day operations may have all come into play. Plus, Pearle Beasley was long past retirement age.

"I don't want it cut up into building lots," Mrs. Beasley (76-years-old at the time) told reporter Jesse Poindexter for the *Winston-Salem Journal*. "It is a beautiful mountain, and I want it to stay that way even after I'm dead and gone."[2]

The intention of the Northwest Economic Development Commission's proposal to buy Pilot Mountain was announced in the Winston-Salem Journal on December 19, 1966, a few months after the jailhouse incident (see chapter 6). The plan would use funds from the federal Land and Water Conservation Fund for half the purchase price and the other half would come from private donations and local industry. $1.6 million had been recently given to the state from the federal government, but according to the article in the journal, "Of more immediate concern is whether the commission could get enough money out of the Land and Water Conservation Act when so many of the state's local governments, the Wildlife Resources Commission and the Piedmont Triad Committee are bidding for funds."[3]

1. Stewart and Roberson, *Exploring the Geology*, 135.
2. Jesse Poindexter, "Pilot Mountain Offered for Sale," *Winston-Salem Journal*, December 17, 1966, Forsyth County Library, Winston-Salem, North Carolina.
3. Staff writer, "Buying Pilot Mountain," *Winston-Salem Journal*, December 19, 1966.

FIGURE 7.3. This image of Joe C. Matthews might have been a bit of a joke. Since reading about what he did in his life, it would seem he was rarely at his desk with his feet up. Courtesy of Joe C. Matthews archives.

For years, much of the land immediately surrounding Pilot Mountain has remained in its natural state – not unlike an untouched wilderness. Here and there are some fields or open pasture land, but by and large it is all rather rough. And when you undertake to fix corners and establish boundaries for 40 or 50 tracts of this land, all under different ownership, and many of which have been in the same family for generations with only ancient deeds to go by, it can turn into quite a task.

That is what Joe C. Matthews, executive director of the Northwest Economic Development Commission, found out when he started to determine the metes and bounds of the proposed Pilot Mountain State Park. . . For several weeks, Matthews has been spending much of his time with a surveyor, attempting to fix the corners and boundaries of the property involved. After tramping over much of the area, talking with numerous owners, consulting Surry County tax maps, and studying old deeds, Mathews is ready to concede that the job is far too much for one man and one surveyor.

—Sherman Shore (1967)[4]

4. Sherman Shore, "Park Limits Hard to Find," *Twin City Sentinel*, hand dated December 23, 1967, Mount Airy Museum Historical Room.

For the next several years, Joe Matthews and a team of volunteers worked tirelessly on the Pilot Mountain project even after it was completed. Once Pilot Mountain State Park was finally a reality, Matthews stayed involved. Matthews's work to save Pilot Mountain has come to fruition with over a million visitors in 2020. Honestly, Pilot Mountain State Park should have a "Joe Matthews Day" out of respect for his herculean efforts to transform the mountain into a state park when many others had been unsuccessful in the previous decades.

North Carolina HISTORIC SITES **STAFF BULLETIN**

VOLUME XIII, NO. 12 DECEMBER 1997

REMEMBERING JOE CARROLL MATTHEWS
April 28, 1932 – November 14, 1997

The staff of Horne Creek Living Historical Farm was profoundly saddened by the death of Joe Carroll Matthews on November 14. With his passing, the northwest piedmont lost a dedicated public servant, the North Carolina Historic Sites Section lost a loyal friend, and our site had to say good-bye to the person deservedly recognized as the "father of Horne Creek Farm."

Who was Joe Matthews? He was a very, very special man. After graduating from the University of North Carolina at Chapel Hill in 1954, he began working for Yadkin County, initially as a teacher, then later as Director of Social Services. In 1966 he became the executive director of what is now the Northwest Piedmont Council of Governments. His accomplishments in that capacity were as extensive as they were varied. Many, however, reflected his great love of the land and its people. Community service work in Yadkin County on a large scale, the development of Pilot Mountain State Park and the Yadkin Trail System, the expansion of Hanging Rock State Park, the creation of Horne Creek Living Historical Farm, the development of Jockey's Ridge State Park, preservation projects including Richmond Hill Law School and Historic Rockford, and the addition of the New River to the National Wild and Scenic River System are simply a few of the projects to which he devoted enormous amounts of time and energy. It was not surprising to those who knew him that he received the Governor's Award for Conservation of Natural Areas in 1977.

Mr. Matthews believed deeply that the state needed a farm museum. When he saw a need for a project, his tenacity in bringing it to fruition was little short of amazing. From before the site's establishment in 1987 to the fall of this year, he remained one of Horne Creek Farm's greatest and most active supporters. From urging state legislators to provide more support for our project to arriving at the site with chain saw in hand ready to help clear away dead trees, he could always be counted upon. Joe loved the project and never asked anything in return, other than that it prosper.

In the words of one director of the North Carolina Living Historical Farm Committee, "Joe leaves a lot of footsteps." They are footsteps of a friend who will be sorely missed.

(Lisa Turney)

Published periodically by North Carolina Historic Sites, a program of the Division of Archives and History, N.C. Department of Cultural Resources for section personnel. The section's home office is at 532 N. Wilmington Street, Raleigh, N.C. 27604. (919) 733-7862. James R. McPherson, Administrator.

FIGURE 7.4. A remembrance of the life of Joe C. Matthews. Courtesy of Lisa Turney.

> Before the meeting, the committee and guests ate a chopped barbecue and fried chicken super [*sic*]. It marked a first – probably the 'highest-up' meal that caterer Paul Myers has ever handled. While the supper was being set up, some of the committee members took the walk to the top of the knob. Others, like Forsyth Sheriff and Mrs. Ernie G. Shore, satisfied themselves with a look from the lower part of the mountaintop. 'That's enough for me right now,' said Shore.
>
> —Bill East (1967)[5]

A forty-one-member committee called the Pilot Mountain Preservation and Park Committee was formed to explore buying the 1,066 acres and to raise funds needed to make Pilot Mountain a public resource. Six weeks later on August 25, 1967, an "Option to Purchase" (OTP) including Pilot Mountain was drawn up and $10 in earnest money was provided. The purchase price agreed upon was $682,500. The following was written into the OTP as a guarantee it would not be used for commercial purposes:

> Mrs. Beasley owns a tract of approximately 1,066 acres in Surry County, North Carolina on which is the distinctive mountain known as 'The Pilot Mountain,' (which tract may be sometimes herein referred to as The Pilot Mountain); and Mrs. Beasley is willing to sell The Pilot Mountain to an appropriate organization or agency which will develop and maintain the property as a park for the use and enjoyment of the public as a State Park; and Park Committee is a non-profit corporation which has been incorporated primarily for the purpose of raising funds for the purchase of The Pilot Mountain, and for such additional property as may be desired, for the purpose of establishing and developing a public park.
>
> —Jesse Poindexter (1966)[6]

CAMPAIGN KICKOFF TO BUY PILOT MOUNTAIN IN 1967

A letter kicked off the campaign to help buy Pilot Mountain with matching funds from the United States Federal Government. The main issue was to preserve Pilot Mountain from commercialization.

5. Bill East, "Meeting on Mountaintop Is Cool," *Winston-Salem Journal*, July 12, 1967.
6. Option To Purchase Real Estate legal form for Pilot Mountain sale, August 25, 1967, between Mrs. Pearle Beasley and Pilot Mountain Preservation and Park Committee, Inc.

> Preservation of the mountain's natural beauty is one of the guarantees made to Mrs. Pearle Beasley by Joe C. Matthews, director of the economic development groups.
>
> —Staff Writer Winston-Salem Journal (1967)[7]

This wasn't the first-time fireworks went off on top of Pilot Mountain. Joseph (Joe) Wallace King had also done a fireworks display a decade earlier on Saturday, April 27, 1957, to announce the Arts Follies in the area.[8] To help with the smoke signals in 1967, King first put an ad in the paper, but after getting no response, he drove up to Cherokee, NC, himself to get some help. There he enlisted the help of Cherokee Chief Walter Jackson, Cherokee Braves David Arch, and Bennett Owl.

> "Joe King, Winston-Salem artist who is doing the stage-setting for the occasion, including a big fireworks display from the mountain top, is throwing in the smoke signaling for good measure.," wrote the Twin City Sentinel newspaper on Saturday, September 23, 1967.

FIGURE 7.5. Image of the blanket after it was used to do smoke signals to kick off the fundraising campaign. Courtesy of the Frank Jones Collection at the Forsyth Public Library.

7. Poindexter, "Pilot Mountain Offered for Sale."
8. Staff writer, "Heap Smoke and Fire Say: Bring Wampum," *Winston-Salem Journal*, April 21, 1957.

For several days, Joe has been running a classified ad in the Journal and Sentinel asking for a couple of full-blooded Indians ('we'll supply the smoke') to perform this chore. Walter Jackson, young chief of the North Carolina Cherokees at Cherokee, and a couple of his braves, David Arch and Bennett Owle, responded to the call... They will leave Asheville early Wednesday morning on a Piedmont Airlines plane scheduled to arrive at Smith Reynolds Airport at 9:08 a.m. They will be guest of King and other campaign leaders during the day... The smoke signals will begin at 6 p.m., while there still is plenty of daylight to make them visible over a wide area.

—Sherman Shore (1967)[9]

"We don't know one smoke signal from another," Owl admitted to writer Bill Wright.[10] "Smoke signaling is a lost art among the Cherokees," confirmed Chief Jackson to Wright to much laughter. Given the wet weather conditions, just starting a fire became an issue until the National Guard stepped in with a large "smoke pot," which worked so well that the signal blanket caught on fire.

"Save our mountain. We want wampum," is what King jokingly wanted them to signal that first night, according to Wright. The delay seemed to add to the excitement, with each day having free updates on the status of the event on the area radio and in newspapers. When it finally did happen, after three days of being postponed, it is estimated over a half million people witnessed it, jamming roads throughout the area in all directions.

Take this script, Indians sending smoke signals from the mountain summit... sky divers plunging toward the mountain with torches, smoke trails and multi-colored parachutes... hundreds of railroad fuses, ringed about the mountain top, bursting into flame... circling planes dropping brilliant flares... and from the very pinnacle of the Pilot, 2,440 feet up in the darkness, burst after burst of fireworks for 45 minutes... theme music, 'The Ballad of Pilot Mountain,' especially written for the occasion. An extravaganza worthy of DeMille.

—Bill Wright (1967)[11]

King announced the Cherokee smoke signal event for a Wednesday night in 1967 to start fundraising for buying the mountain, but rain delayed it to the following Saturday night. The headline for the Winston-Salem Journal the next day

9. Sherman Shore, "Progress Reports: Delay Feared in Plans for Park," *Twin City Sentinel*, September 23, 1967.
10. Bill Wright, "From Murphy to Manteo," *The State*, November 15,1967.
11. Wright, "From Murphy to Manteo."

was, "RJR Gives $100,000: Fireworks Fizzle in The Surry Drizzle."[12] Karen McAdams, whose home at the time was so close to Pilot Mountain that it is now within the State Park, recalls this event:

> I do remember the fireworks off the top of Pilot Mountain. I guess that I was 11 years old if it was in 1967. It was very exciting for a kid. I vaguely remember smoke signals off the top, but can't remember if that was the same time or not. Pilot Mountain was always beautiful in the snow, or in the fall with colorful leaves, or in the spring with the different shades of green on the trees as the leaves came out. It is really beautiful all the time.
>
> —Karen McAdams (2020)[13]

It was noted in the advance newspaper features that forty to fifty firemen would be on hand for the fireworks and they would linger long after it was all over to make sure sparks from the fireworks didn't create a fire on the mountain. The fireworks were to go off over five hundred feet above the mountain top to ensure that a fire wouldn't be started by an accident below on the slopes or surrounding areas.

Joseph Wallace King, a member of the Pilot Mountain Preservation and Park Committee, organized the event. King arranged for twenty-five volunteers to

FIGURE 7.6. Karen McAdams' barn in Mebane, North Carolina, painted by her daughter but unfinished at the time this image was taken.

12. Jesse Poindexter, "RJR Gives $100,000: Fireworks Fizzle in Surry Drizzle," *Winston-Salem Journal*, September 28, 1967.
13. Karen McAdams, email correspondence with author, September 10, 2020.

FIGURE 7.7. View from the Pilot Mountain Country Club in 2022, where the volunteers watched the fireworks in 1967.

scale the 106 steps up to the Big Pinnacle and then light five hundred railroad fuses around the top of Pilot Mountain. Tony Vitale, of New Castle, PA, designed the fireworks display. Lloyds of London insured the event for $500,000.

> Cars lined every road from Winston-Salem to Mount Airy, and were bumper-to-bumper in both lanes of dual Highway 52, both north and south. Traffic was further clogged, reported Mount Airy News, by crowds from King and Mount Airy fairs. It took some motorists three hours to travel the 25 miles from Pilot Mountain to Winston-Salem after the display.
>
> —Bill Wright (1967)[14]

Huber Hanes Jr., a textile executive from Winston-Salem, hosted a BBQ for two hundred campaign workers and the committee at the Pilot Mountain Country Club before the fireworks just northeast of the mountain. This is where seventeen-year-old Lucy Carol Davis debuted "The Ballad of Pilot Mountain."

THE "BALLAD OF PILOT MOUNTAIN" SONG WAS A TEAM EFFORT

The original lyrics for "The Ballad of Pilot Mountain" were written for a Public Service Announcement (PSA) ad campaign by the Long, Haymes & Carr public

14. Wright, "From Murphy to Manteo."

relations agency in Winston-Salem for the opening of the Pilot Mountain fund-raising drive. They sent the lyrics to the North Carolina School of The Arts in Winston-Salem, requesting a student put music to it. Seventeen-year-old Lucy Carol Davis did come up with the music and added a few lyrics of her own. That was when her college life changed and she had a run of fame on campus as the PSA she filmed aired over and over again to announce the fundraiser to buy Pilot Mountain.

The song starts by describing the mountain, how unique it is, and asking ten thousand people to preserve the mountain's fame.

> The song goes on to tell how Indians and Daniel Boone used it for a landmark and how it was surveyed in 1751 by the father of President Thomas Jefferson. Then the ballad asks for help to buy the mountain and make it a state park.
>
> —Jeanette Reid (1967)[15]

The week of the fireworks, a helicopter was dispatched to pick up Davis at her school and landed on the athletic field to pick her up. That landing got a good bit of attention at the school. Page Shamburger, journalist for Rotor & Wing Magazine of Professional Helicopter Operations, wrote about the filming of the PSA in detail. The helicopter landed with them there at the makeshift airfield area on top of the mountain that both W. L. Spoon and J. W. Beasley had hoped would be an airfield for just such activity.

Once on Pilot Mountain, after that brief helicopter ride, Davis and a film crew had to get all their equipment from the parking lot to the filming location. That meant going across the saddle area and then up the wooden stairway to the top of the Big Pinnacle by foot. The helicopter also filmed zooming in and out with a second camera for the commercial. Tom Hilderbrand from H & D Productions, Southern Pines, told Shamburger:

> For instance, this last film I made of a folk singer, Lucy Carol Davis, on top of Pilot Mountain – we got the singer, the mountain, and the whole feeling of space from the 'copter. Without the 'copter, we could not have made the shot, and without that shot, the film would've lacked a great deal.
>
> The Pilot Mountain film is an excellent example. The peak, a lone sentry several miles east of the Smokies chain, stirs the nearby air. This turbulence shook the Hughes [Helicopter] too much for the camera to

15. Jeanette Reid, "Balladeer Sings for a State Park at Pilot Mountain," *Winston-Salem Journal*, ca. 1967, Davie County Public Library, Mocksville, NC.

> chronicle the singer fading out in the distance, so Tom inverted his camera and took the fade-out going in. Result? A beautiful smooth view of the singer strumming her guitar as she disappeared into the distant horizon, and who would know the film was shot backwards?
>
> —Page Shamburger (1967)[16]

Davis still has a VHS tape of the PSA, but it falls well behind today's HD standards. The helicopter came and went for three days, giving the Beasley family members a ride around the mountain for an even more spectacular view of Pilot Mountain from the air.

FUNDRAISING CAMPAIGN COMPLETED IN JUNE OF 1968

The fundraising goal was for ten thousand people to contribute. That goal was exceeded when more than twelve thousand people contributed $370,000. When combined with $1,040,000 in federal funds, it was enough to buy the 1,066 acres from the Beasley family, plus the acreage around the mountain. The money was all combined, purchasing 2,145 acres for Pilot Mountain State Park. While the State of NC approved the state park, they did not contribute any funds to the purchase of the mountain. This was absolutely brilliant work by Joe Matthews and his team of volunteers. Lucy Carol Davis also confirmed that Andy Griffith attended a few of the fundraising events to buy Pilot Mountain.

PILOT MOUNTAIN STATE PARK DOESN'T OWN ALL OF THE MOUNTAIN

As noted in the last chapter, the deed of sale had to exempt the acre on top of Pilot Mountain owned by the Western Union and Telegraph. They had bought that acre from J. W. Beasley on September 22, 1947. This agreement was iron-clad, thwarting Pearle Beasley from buying it back.

There were three sales agreements during the fundraising effort, and the property was only agreed to be for a park in the sale. The final deed made the 1,066 acres the Pilot Mountain State Park, but it also included a total of thirty-five tracts of land. Some of those tracts of land were crossed out within the deed for unknown reasons.

On June 24, 1968, at 3:45 p.m., Thomas C. Ellis represented the State of NC at the Surry County Courthouse signing. He handed over the check for $682,500 to Pearle Beasley for the entirety of Pilot Mountain except for that one acre. Shortly after, a phone call was made to the toll booth at Pilot Mountain

---

16. Page Shamburger, "A Better Mousetrap!" *Rotor & Wing Magazine*, March 1968, 10–13, 44.

informing them that the per-car admission price was waived from now on.[17] Pilot Mountain State Park was finally a reality! In 2021, the Pilot Mountain State Park tax appraisal card was valued at $11,079,060.[18]

The pool was retired, and then the lower flight of stairs to the Big Pinnacle were removed as the State Park service took control of the property. Rock climbing on the Big Pinnacle also came to an end. However, rock climbing is still allowed on the Ledge Spring Trail and, on weekends, it often has an amazing number of people on the rock faces.

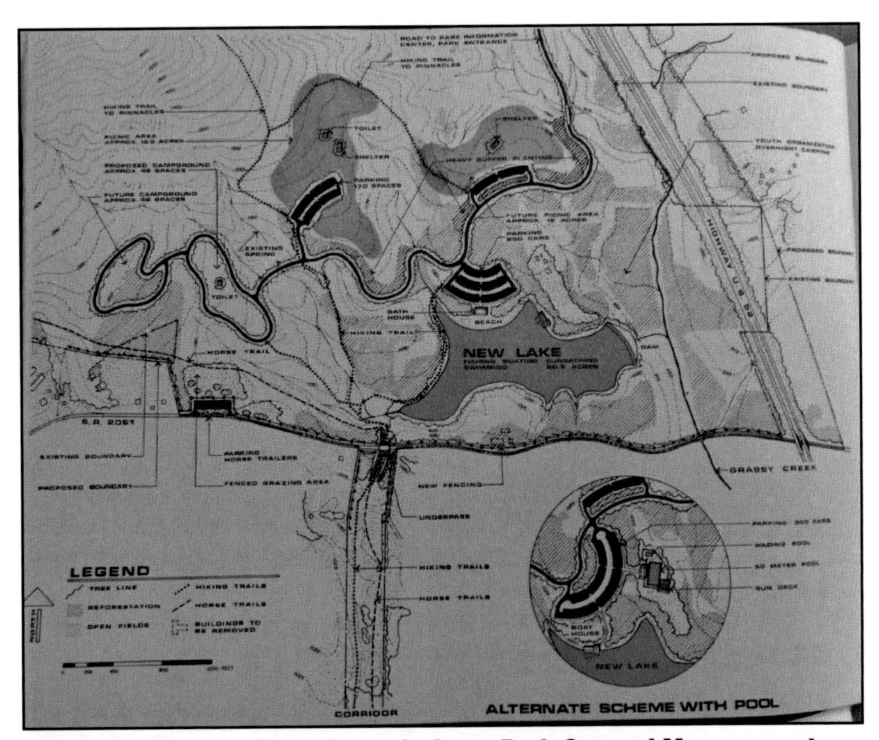

FIGURE 7.8. For the Pilot Mountain State Park General Management Plan of 1970, new trails, a new fifty-meter pool, a sun deck, a 12-acre picnic area that was near a 20.5-acre man-made lake with extensive campgrounds, and hundreds of parking spaces were all in planning stages. Had they built the planned fencing around the property, that would have been a time to have "contract archeology." Once and for all, that could have determined if there was a Mound Culture palisade with a moat around the mountain in the prehistoric past. The next book delves into that aspect of the mountain in terms of a nearly perfect earthwork pyramid, palisade, moat, and a possible gigantic longhouse where the picnic area is now.

17. Staff writer, "State Park," ca. June 1968, Forsyth Public Library, Winston-Salem, North Carolina.
18. Surry County Tax Office, Parcel ID: 5964-00-46-5352, checked on October 10, 2022, by an Assessment & Collections Technician, Dobson, NC 27017.

STEPS UP TO THE BIG PINNACLE ON PILOT MOUNTAIN CONDEMNED

> Pilot Mountain (AP) – The steps up Pilot Mountain's famous rock knob have been condemned by state inspectors and blocked off to keep people from climbing them. A sign has been posted at the bottom of the steps warning that they are not safe, and the first short segment of steps has been removed to make the rest inaccessible. But, Carl Ray Flinchum, the superintendent of Pilot Mountain State Park, said yesterday that really determined people are ignoring all these things and scrambling up several feet of rock in order to climb the rickety steps that remain. New steps up the pinnacle were listed as one of the park's most pressing needs when the 1,000-acre mountain and approximately 1,100 acres around it became a state park in July 1968. . . Funds for the steps are one of the park's department's budget requests for the 1971-73 biennium.
>
> —Arlene Edwards (1970)[19]

The issue of the stairs to the top of the Big Pinnacle remained a hot topic for the public with several news items and some strongly worded letters to the editor on the matter.

RAVENS BECOME PART OF THE ISSUE FOR STAIRS ON THE BIG PINNACLE

> Many ravens were constantly flying about the pinnacle; they seem to use this place for building their nests and rearing their young; but the only permanent inhabitant I found at this elevated spot, were three rabbits taken up some six months ago by some mischievous boys; and how they live here without water, is a problem that I cannot solve.
>
> —Publius (1849)[20]

Ravens were found once again nesting on the east side of the Big Pinnacle. "Pilot Mountain Now Haven for the Raven" read the 1977 new headline.[21] Clearly, the ravens were not a new development and had been there for at least the last couple

---

19. Arlene Edwards, "41 Years Old: Mountain Steps Condemned," *Winston-Salem Journal*, July 2, 1970.
20. Publius, "A Trip to the Mountains," *The Greensboro Patriot*, July 28, 1849.
21. Bob Simpson, "Pilot Mountain Now Haven for the Raven," *The News & Observer*, May 29, 1977.

of centuries with daily human traffic on top of the Big Pinnacle near the nest. During that long time period, the Big Pinnacle hosted thousands of excursions to the top. The state started taking their responsibility to protect Pilot Mountain seriously and enlisted a group of volunteer experts including botanists, biologists, zoologists, and ornithologists. It was decided not to reopen the stairs to the top of the Big Pinnacle to the public.[22] Money had already been allocated to rebuild the stairs, but the impact statement halted this effort. It was put forth that the hermit-like ravens would leave if people were to tread near their nests. This cliff nesting spot is on the far east side of the Big Pinnacle on the opposite side of the Big Pinnacle from the stairs. This cliff location would only be accessible to rock climbers. In addition, four species of rare plants could be endangered by heavy public use but it has been decades since any survey could be found. Rebuilt stairs at the same location they have always been located on the west side of the Big Pinnacle would keep humans away from the ravens. A small raised platform on the top would be a win-win for all concerned in terms of the unique plant life found there on the Big Pinnacle.

On 22 March 1974 a team comprised of Fran Baldwin, Ruth Hill, Jerry Shiffert, Ramona Snavely, Paul Spain, and Park Ranger Jonathan Wild confirmed the nesting of Common Ravens (Corvus corax) on the pinnacle of Pilot Mountain (elevation, 2,440 feet) in Surry County, N.C. The nest was located on the east side of the pinnacle on a rock ledge with a larger rock ledge diagonally overhanging the nest on the left side. It is quite inaccessible and very difficult to reach.

The nest was a large bulky structure about 16 to 20 inches outside diameter and about 10 to 12 inches inside diameter consisting of large twigs l/4 to 1/2 inch in diameter. The young could not be seen in the nest because observation had to be made from below rather than above. Two and possibly three young were heard calling each time the adults approached the nest. The adults were seen carrying food but fed the young only once while we were present. Numerous photographs were taken of the nest, the surrounding habitat, and the adult birds.

A second visit was made on 7 April 1974 by Wayne Irvin and Jonathan Wild. On this occasion two young and possibly a third' one were seen peering above the nest. The adults fed the young, and photographs were taken.

—Ramona R. Snavely (1974)[23]

22. Nash Herndon, "Save Pilot Mountain Ravens, Report Urges: Building of Stairs to Summit Could Endanger Rare Nesting Place," *Twin City Sentinel*, August 17, 1976.
23. Ramona R. Snavely, "Nesting Ravens on Pilot Mountain," Carolina Bird Club, *The Chat* 38, no. 3 (September 1974), 75, https://www.carolinabirdclub.org/chat/issues/1974/v38n3.html.

Despite all this information, and perhaps because of the lack of communication between various involved parties, as late as 1979, bids were taken for new stairs but the budget was $45,000. The bids ranged from $107,532 up to $125,045 to bring them up to modern OSHA standards of safety. Sadly, a child died in a fall trying to scale the Big Pinnacle, so the remaining steps were removed or are hidden from view to this day. According to the Pilot Mountain Rescue, the timber rattlesnakes are still an issue on the Big Pinnacle but are rarely seen on the trails. They warn that if you get off the trails, you should be aware of the timber rattlesnakes and large black racers that are all a vital part of the flora and fauna of Pilot Mountain State Park.

TELL THE MEDIA THE TIGERS ON PILOT MOUNTAIN ARE FOUND

"I went to work my first day on Monday, April the second of 1984," Sherrie Robin Lynch-Bennett (a.k.a. Robin L. Bennett), who is now retired from being an office assistant, recalled of her eventful first days on the job in the Pilot Mountain State Park office. She continues:

> Larry [Melton], the park superintendent, left that afternoon about three or three thirty. He was going to a meeting that they were having. Well, after he left, it was raining and the fog moving in so I went home at five. When I went in on Tuesday morning, I drive in the parking lot at the office and it's just lots of cars. I could tell that the Search and Rescue Squad was there. Right after I got in the office, you know, we use the radio to communicate with the Rangers and the Superintendent. 'If anybody from the media calls tell them the tigers have been found,' Larry calmly said over the radio from the top of the mountain to the office (or, at least, that is the way the new employee heard it on her second day of work). Then in a few minutes, the Mt. Airy News called to ask if 'there was any news on the missing hikers that were lost in the park last night?' My gosh lady, if you just knew what I understood about that, I thought, but I didn't tell her. There was a lot of fog that moved in and you couldn't see the knob from the office. You weren't supposed to go up on the knob. After they got up on top of the knob, the fog moving in and they could not see how to get back down... They stayed up on the knob all night long and it was still cold. Oh my gosh! They stayed in a cave, but there's no cave. I was told there was no cave up on top of the mountain. There's like a, you know, an overhang or something but it's not a cave, right?
>
> —Robin L. Bennett (2021)[24]

24. Sherrie Robin Lynch-Bennett, phone interview with author, 2021.

FIGURE 7.9. Some of the many stone steps air-lifted to Pilot Mountain State Park by the NC National Guard as a training exercise. This was a huge win-win for visitors to Pilot Mountain State Park today.

The hikers were found just fine after a night on the top of the Big Pinnacle. They claimed they spent the night in a cave out of the fog. Another interesting report of strange occurrences on Pilot Mountain, since there isn't a cave up there.

NATIONAL GUARD BLACKHAWK HELICOPTERS ALL OVER PILOT MOUNTAIN

Another eventful day happened the very first morning when improvements for the Pilot Knob Trail were undertaken on the Big Pinnacle of Pilot Mountain State Park. Up until that time, the trail had been the natural one for centuries but was judged to be too rough for the general public. Frederica Lashley, the president of The Unturned Stone Inc., got the contract.

> I remember when the Jomeokee trail [a.k.a. Pilot Knob Trail] was restored and refurbished, A lady contractor got that job and I think she was out of Lexington [NC]. She had no way to get the material up there on the trail so they got some Blackhawk helicopters to do it. They had all the material delivered to the summit parking lot. Then the helicopters would glide it around and drop it where they were told. Both the Winston-Salem Journal and the Mt. Airy News put that in the paper. She was an excellent person working on that.
>
> —Robin L. Bennett (2021)[25]

25. Bennett, phone interview.

Over the course of a few weeks, the Air National Guard UH-60L Blackhawk helicopters were able to deliver over four-hundred tons of gravel and two-hundred tons of large stones that were made into steps at eight different locations on the Pilot Knob Trail around the Big Pinnacle. The helicopter had a 120-foot line and usually had one or two tons of stone at a time to place on the trail, often right between a stand of trees. This turned out to be a win-win for the park service and the Air National Guard. The Pilot Mountain State Park saved a lot of money, and the Air National Guard got to practice loading and unloading under interesting circumstances. The whole project improving the trail was $429,000, but they saved $250,000 in additional expense thanks to the Air National Guard using it as a training exercise. The first day, the Blackhawk helicopters caused a stir in the Pilot Mountain community and at the Pilot Mountain State Park office.

> Once the helicopter got to going, the phone got to ringing, and it kept ringing. About 30 people called and 20 came to the office to ask about the helicopter Wednesday, the first day of airlifts. Robert Nickell stopped at the office with his 4-year-old son, Joseph. 'I thought they were probably rescuing someone who was climbing up on the ledge,' Nickell said. I had numerous people to come and tell us, you know, how improved it was.
>
> —Staff Writer Winston-Salem Journal (1984)[26]

> When it first started there were some people that got so upset, because the natural beauty was going to be ruined. They were just upset! They didn't want nature to be bothered, but I was up there one day after work. These people came up through there and they had a person in a wheelchair. Before that paved sidewalk on the upper side of the parking a person like that couldn't go up there at all.
>
> —Robin L. Bennett (2021)[27]

ANDY GRIFFITH AND PILOT MOUNTAIN (A.K.A. MOUNT PILOT)

Pilot Mountain was also known far and wide on early television under a different name. Locals knew just what Andy Griffith was referring to from the get-go in his show. The fictional city of Mount Pilot was frequently mentioned on *The Andy Griffith Show*. Andy Griffith himself was born just a few miles from Pilot

26. Lisa Hoppenjans, "Uplifting Experience – Helicopter Moves Gear for Pilot Mountain Renovation – Guardsmen Get Practice," *Winston-Salem Journal*, February 20, 2004, 1.

27. Bennett, phone interview.

Mountain in real life. Griffith's real life best friend, Earlie Gilley, lived in Pilot Mountain. Earlie Gilley was name-checked on the show several times as Earl Gilley. Even though Griffith insisted his show was fictional and not based on anything real, the Gilley story contradicts that. In addition, the Snappy Lunch restaurant has been in operation since 1923 in the town of Mount Airy, which was also part of the fictional show. The one wall of the sheriff's home living room was a granite fireplace and that certainly is a feature of the town of Mount Airy.

The real-life Gilley did have an auto repair shop just northeast of the city of Pilot Mountain. The most memorable episode that referenced Earl Gilley was the one where he didn't caulk his boat and it sank while he was out fishing on a lake for a fishing contest early in the series. The real-life Earlie Gilley has kept the original road grader that W. L. Spoon used in 1929 to help build the clay and sand road to the top of Pilot Mountain all these years later. The Gilley family have attempted to donate the Spoon road grader to Pilot Mountain State Park, but since there wasn't any interest, it is often cleaned up, painted, and put on display with lights at Christmastime. Earlie Gilley married Lorraine Beasley in 1946. Her name, Beasley, was name-checked on the show as the last name for Barney Fife's girlfriend.

> Gilley's Automotive. 'Goober's Replacement,' Episode #186. Goober considers moving to Mount Pilot to work for Earlie Gilley at his garage. Until he passed away, the real Earlie Gilley resided in Pilot Mountain, North Carolina, with his wife Lorraine Beasley Gilley (Andy's first cousin). He owned and operated a service station and garage. Earlie is mentioned in four episodes of The Andy Griffith Show. Lorraine always said, 'Andy just loved Earlie's name!'
>
> —Andy Griffith Museum in Mt. Airy, North Carolina (2021)[28]

A story that circulates around Pilot Mountain is that Griffith's mother wanted him to buy Pilot Mountain from the Beasley family well before it become the state park it is today. In 2020, I was able to confirm this rumor in a telephone interview with Patricia Sperry, Pearle Beasley's granddaughter. Sperry mentioned he wanted to have a Matterhorn style ride around the Big Pinnacle. This would have been a few years after Disneyland debuted their "Matterhorn flying bobsleds" in June 14, 1959. Winifred Coleman also confirmed that Andy Griffith wanted to buy Pilot Mountain and have a cable car go up to the top.

28. Plaque on wall of Andy Griffith Museum, 218 Rockford St, Mount Airy, NC 27030-4662.

> I do remember waiting one afternoon for Andy Griffith to come to an appointment his mother had made with my Grandmother Pearle Beasley. He never showed up, but I kept going to the front door looking for him to arrive for a meeting arranged by his mother.
>
> —Patricia Sperry (2020)[29]

"Over the years a lot of people have come to believe that Mayberry is based on my hometown and it is not," Griffith was quoted in the *Chicago Tribune*.[30] "Cause real towns have real problems that have to be dealt with. All of Mayberry's problems were solved in half an hour."

Despite this claim by Griffith, Mount Airy has "Mayberry Days" each year on the last weekend of September, which is wildly successful. The mentor for writing this book was Art Fettig, who coincidently looks so much like Andy Griffith that he was the parade marshal for Mayberry Days and had a stand-up comedy show, "Almost Andy." While Mount Airy is certainly a tourist destination, it also still exudes a small-town charm that is undeniable. This is a win-win for the town and the fans of the show.

FIGURE 7.10. If you were a fan of the Andy Griffith Show, like this bunch, you'd be disappointed too that Snappy Lunch and Floyd's City Barber Shop were closed in Mount Airy. Fortunately, Aunt Bea's Restaurant was open just outside of downtown Mount Airy on the Andy Griffith Parkway for that young man in the middle of the image.

29. Patricia Sperry, phone interview with author, 2020.
30. Mark Jacob and Stephan Benzkofer, *Chicago Tribune*, Sunday, April 26, 2015, Section 1, 23.

PAINTER BECOMES FILM DIRECTOR

Referencing his film, *Somebody Moved My Mountain*, Joseph Wallace King's widow insists the buying of the mountain was in reference to Black Mountain and not Pilot Mountain. The opening thirteen minutes of this movie featured stunning stunt flying around the town of Pinnacle and Pilot Mountain State Park itself. There is a beautiful tip of the hat to the Cherokee experience, specifically about Pilot Mountain, in the dialogue filmed on the Little Pinnacle before the safety railing was installed. The film came out when independent films were very popular. Since the film featured western North Carolina and had prominent individuals as extras whenever it was shown, it was very popular and well attended.

FILM TRIBUTE TO STUNT FLYING AROUND PILOT MOUNTAIN

While it is likely Zach Reynolds, R. J. Reynolds' grandson, had a life-long dream to be a movie stunt man, his flying was not filmed by King for his independent feature film. However, Reynolds might have been the inspiration for the stunt flying in King's movie. Reynolds did the dangerous stunt or aerobatic flying at air shows and county fairs. Throughout the 1974 movie, you can see stunt flyer Jimmy Dean's distinctive yellow S1-C Flatwing Pitts Special airplane. Reynolds had the exact same Pitts Special airplane in bright red. We know that by 1974, Reynolds had sworn off stunt flying due his recent marriage and a vivid dream he had that he would die in a plane accident. Sadly, that tragic dream did come true when he and three others crashed within sight of Pilot Mountain in a wooded area in Pinnacle, NC.[31] Reynolds broke a pact with himself to not fly anymore after he got married. A neighbor got his pilot's license and Reynolds agreed to fly in the plane as the copilot.

> Initially he (Joe King) wanted me to fly under that bridge and I was going to do it. At the last minute he changed his mind and said to just land and taxi under it, either way it remained sort of outlandish so as soon as I got up to it, I just put the fire to it and pulled it back around. Then I landed before the bridge and taxied up the off ramp and shut the airplane down. Went out and talked to him about what we were going to do. There was a lot on Pilot Mountain that was cut out of the film but I guess that always happens. One day there were tons of buzzards. There is a lot of air currents, thermals and stuff like that.
>
> —Jimmy Dean(2021)[32]

31. Associated Press, "Reynolds Had Dreams of Death," *The Charlotte Observer*, September 28, 1979.
32. Jimmy Dean, phone interview with author, 2021.

LESTER FLATT'S MOUNT PILOT BLUEGRASS FESTIVAL

In 1974, the legendary bluegrass picker Lester Flatt, of the bluegrass band Flatt & Scruggs, bought a large campground at the base of Pilot Mountain, southast in Pinnacle.[33] Flatt, Arthur Smith (who had a music variety show in Charlotte where a young Charles Kuralt got his start), and the Stoneman Family all at different times wanted to hold a bluegrass festival on top of Pilot Mountain in the huge parking lot and airport area, but those ideas were turned down by Pearle Beasley.

In the 1970s, Flatt held a well-attended Mount Pilot Bluegrass Festival, attracting thousands of music fans to a huge natural outdoor amphitheater. Clearly, the name of the festival was a tip of the hat to Andy Griffith, whose show used the Dillard's bluegrass band as guests. The main stage and the smaller Lester Flatt's Pickin' Stage for the annual festival both feature the unmistakable southern exposure of Pilot Mountain looming large in the background. Flatt bought the Jomeokee Campground from the previous owner, Tom Jones. Flatt's manager announced they would turn it into a seasonal campground and improve the entertainment facility. The natural amphitheater can hold twenty-thousand people, according to the current owner, Thomas Pace.

Lester Flatt's Mount Pilot Bluegrass Festival featured national bluegrass stars and drew large crowds. Lester Flatt & the Nashville Grass headlined each

FIGURE 7.11. Mainstage for natural amphitheater at Lester Flatt's Mount Pilot Bluegrass Festival. Was this natural area used in prehistoric times like a huge festival for the Green Corn Ceremony? This venue was mainly used in the cool of the evening for concerts during the festival. Honestly, this location rivals the Red Rocks amphitheater in Colorado in terms of the view in a natural setting.

33. Staff writer, "Lester Flatt Buys Jomeokee," *The High Point Enterprise*, November 17, 1974.

FIGURE 7.12. Daytime shaded Pickin' Stage at the Jomeokee Campground, where Lester Flatt held bluegrass festivals in the 1970s. This image was taken in 2022 with a wonderful view of Pilot Mountain in the background.

festival, which in 1974, also included Bill Monroe & the Bluegrass Boys, Mother Maybelle & the Carter Family, Jimmy Martin & the Sunny Mountain Boys, Grandpa Jones and Romona, and Benny Martin, among others. A coveted prize, the winner of the amateur fiddle contest at noon on Saturday would appear on WSM's Grand Old Opry and get a tour of Opryland.[34]

FIRE ON PILOT MOUNTAIN AGAIN AND AGAIN

Stories of fires on Pilot Mountain go all the way back to prehistoric times. There has even been a thesis paper written on the Pilot Mountain fires, *Fire in central Piedmont as recorded by fire scars at Pilot Mountain State Park, NC,* by Dane Mitchell Kuppinger and Abigail Rich, which was noted in the Spoon chapter.[35] Local news records of the Pilot Mountain fires date as far back as one hundred and fifty years. There were even fires on Pilot Mountain in 2012 and again in 2021 during the writing of this book.

> The park's master plan written at its founding says that it was the local custom to set fires on the mountain at Eastertime until the mid-1930s and that the last 'significant fire occurred in 1927. Two 'small' fires were reported between 1929 and 1944, but their extent, severity, and exact

34. "Lester Flatt's Mount Pilot Bluegrass Festival," *Statesville Record and Landmark,* June 22,1974.
35. Kuppinger and Rich, "Fire in Central Piedmont."

FIGURE 7.13. Fire engulfed the entire mountain within twenty-three hours in November of 2021, but it quickly transitioned into a control burn.

date were not recorded (Williams & Oosting, 1944). Between 1948 and 1969, there were two lightning-ignited and four human-ignited fires. The largest of these, in 1960 and 1961, were set by hunters and burned 1 and 5.7 ha (3 and 14 acres), respectively (Bell & Morse, 1970). The NC Park Service began using prescribed fire within the park in 2009 and since then the entire upland portion of the park has been burned at least once (Windsor, pers. Comm.).

—Dane Mitchell Kuppinger & Abigail Rich (2019)[36]

'Pilot Mountain Aflame': A forest fire raged all day yesterday on famous old Pilot Mountain. The fire, the origin of which is unknown, broke out Tuesday night and was only gotten under control after burning twenty-four hours. The flames did the greatest damage on the west side of the mountain and extended over from 500 to 600 acres. The sight was spectacular, the flames leaping up the sides of the rocky pinnacle that is so well known as a landmark. The house and stable belonging to Mr. N. J. Spainhour was destroyed.

—Staff Writer (1904)[37]

36. Kuppinger and Rich, "Fire in Central Piedmont."
37. Staff writer, "Pilot Mountain Aflame," *Winston-Salem Journal*, April 21, 1904.

Once the road leading to the top of Pilot Mountain was completed in 1929, the fires were easier to contain, resulting in smaller fires totaling a few acres instead of hundreds.

> 'Fire has been kept off Pilot Mountain for three years,' said owner W. L. Spoon of Alamance County yesterday. And now the effect of the good work can be seen; the mountain is beautiful with its wealth of rhododendron and mountain laurel. For years the mountain was neglected. Drinking parties were held there, fires broke out and the mountainsides with all their rich possibilities were burned off regularly. State Forester J. S. Holmes, one of the most efficient men in his line in the country, was appealed to. Some of the State fire patrolmen were assigned the task of keeping an eye on the mountain, and now, as I say, the results can be seen. It is a beauty lesson for the whole State. It proves what I have always contended: that it pays to preserve the beauty of North Carolina.
>
> —Staff Writer (1932)[38]

In 2012, a small control burn got out of hand and ended up burning six hundred acres on Pilot Mountain. Possibly due to social media, this fire received the greatest news coverage, at least until the 2021 fire. After the 2012 fire had gotten out of control, the Pilot Mountain State Park superintendent, Matt Windsor, and Winston-Salem Journal reporter David Rolfe reflected on the damage to the mountain on May 4, 2013:

> Seventy years of fire suppression is not letting nature take its course, we have to use prescribed fire as a tool, and it is as unnatural to withhold fire from a forest in the Southeast as it would be to withhold water. The fire cleared out decades of accumulated debris and under story, breaking holes in the forest canopy overhead, allowing sunlight and clearing the ground for different plant species to take hold... Rain came on Monday, Nov. 12, helping tamp the fire down. Three days later, the fire was declared contained. Windsor said wildlife on the mountain appeared unconcerned by the wildfire, appearing instead to take advantage of opportunities it presented. Several firefighters told me they saw a buck and doe bed down by the fire, get settled in and move downhill when the fire when it got too close, I saw wild turkeys standing at the edge of the fire, running in as soon as the fire had passed to eat beetles and worms driven out by the heat. The turkeys were actually following the bulldozer. Fire is a way for forests to regenerate. By keeping a rein on periodic fires, forests become homogenous and less resilient. The larger trees, like the

38. Staff writer, "Fires and Drinking Kept off Pilot Mountain," *The News & Observer*, May 28, 1932.

chestnut oaks, red oaks, black oaks, will return stronger, and the pitch pines are seeding an entire new generation.

—David Rolfe (2013)[39]

FIGURES 7.14 and 7.15. These images are of a successful Pilot Mountain State Park control burn. The weather whiteboard information was in 2020 at the Pilot Creek Access near the mountain.

39. David Rolfe, "Rising Through the Ashes," *Winston-Salem Journal*, May 3, 2013, Section A, 1.

MULTIDISCIPLINARY ARCHEOASTRONOMY OF PILOT MOUNTAIN

The weekend of the 2021 fire was an extreme red flag situation because the air was so dry. Fires were discouraged across the state. It was the perfect situation for a campfire to get out of control, and it did. Since that area on the Big Pinnacle lit up about thirty hours after the 2021 fire started, it might be much easier now to find that perfect smoke signal spot that the Native Americans used for centuries, if not thousands of years. Because of its prehistoric use, that spot for the smoke signals could be an archeoastronomy bonanza. Local members of the Cherokee Nation may be able to help those who are curious to see this spot. Elsewhere during the 2021 fire, it was a perfect control burn after the first twenty hours or so. However, the two pinnacles lit up like torches that first Sunday evening about 10:30 p.m. It took months to repair the Little Pinnacle viewing area after the intense fire at that possible cairn spot.

> Just recently we met with the North Carolina Forest Service on our plans to burn the property. So when we got the call there was a wildfire on the property [Saturday, November 27, 2021] we already had boundary lines and kind of a layout how we would manage this fire. When the fire occurred we pretty much had a pre-plan in place for the fire. We were able to take that plan that we had previously established months ago and apply it to manage the fire.
>
> —Paul Craig (2021)[40]

That notch in the Big Pinnacle could have profound importance for archeoastronomy. It could teach us about the ancient significance of the rising and setting of the sun and the placement of the moon and planets at particular times of the year. On the east-west axis of Pilot Mountain that takes in the cairn on the Little Pinnacle, the notch on the Big Pinnacle and the arc of the sun rising on either vernal (spring) or autumnal (fall) equinox is a perfect example of this. But could it all be a natural coincidence? A multipurpose notch in the landscape to note something isn't unique to Pilot Mountain. Man-made notches in the landscape have been discovered across the planet, often associated with point-to-point travel. They have also been used to link sacred sites in a straight line, like the cathedrals of Europe, and sometimes reveal the cosmic calendar just like the one at Pilot Mountain State Park on the equinox.

The impression given over the years of research is that the archeoastronomy aspects of Pilot Mountain are like the various functions of a Swiss Army Knife. Let's explore just one aspect of it a little deeper. Ancient peoples wanted to know

40. Paul Craig, "Grindstone Fire and Sauratown Mountain Community Meeting," NC Forest Service, December 1, 2021, 42:50, https://www.youtube.com/watch?v=oNDU23NO0Xs&t=2570s.

the seasons because they were farmers. Before European colonization, nearly all of what is now the United States was tended like it was one large garden. Fire was used to keep the undergrowth down across this land from sea to shining sea. To get a good idea of this, the book *An Indigenous People's History of the United States* by Roxanne Dunbar-Ortiz is a good place to start. Another one is the popular book *1491* by Charles C. Mann. For a scientific analysis, *The Indigenous Paleolithic of the Western Hemisphere* by Paulette F. C. Steeves is a comprehensive collection of scientific papers, though it is also balanced with Indigenous oral histories.

So, where is that center pole-point on the Big Pinnacle of Pilot Mountain State Park? Even though it is simple to plant a stick to show the sun's movement near your home, it is special to gather as a crowd to see the shadow's alignment in the landscape at a place of significance. In modern times, we gather with astronomy groups to view the heavens together at a specific location away from city lights, but one could just as easily do it at home.

FIGURE 7.16. When this image was taken in 2021, the Pilot Mountain State Park controlled burn was well in effect within twenty-four hours of the wildfire's start. Color images make the orange glowing smoke reflecting the low control burn fires seem like fire was huge and out of control. Those distorted images had an artistic flare for the photographer, but they gave the wrong impression, flooding social media and newspapers in the area. However, the Little and Big Pinnacles area flames did shoot high into the air for a time on that first Sunday night of the week-long fire.

FIGURE 7.17. The possible reason for the wide notch on the center of the Big Pinnacle of Pilot Mountain State Park was revealed on the vernal equinox of 2022. This was revealed because almost all the trees on the Big Pinnacle had been burned during the 2021 fire. Of course, the notch will disappear again as the trees return over time. This image is reminiscent of the 23.5-degree tilt of the earth and the tilt of the airfield on the mountain.

According to researcher Randall Carlson, this is what happens with the sun's orientation to the earth on the equinox. Here, Carlson is talking in general terms, but it still applies to Pilot Mountain State Park during the equinox. The equinox is a very special time of the year for the mountain, since it is on a near perfect east-west axis on the earth.

> On the equinox the sun rises, more or less, due east. It is always going to be a little bit off depending on your latitude... of course depending on the day cycle you are nine hours later when it's setting. The earth has already moved. Now this is going to be from the perspective of the northern hemisphere, the rising and setting positions of the sun on the horizon are going to be shifting further and further south until they reach the solstice. Once they reach the solstice then it pauses there for a few days, or appears to anyway, and then it reverses direction and comes back. As it approaches that point the speed with which it is moving is slowing down the whole time until it comes finally to a stop, hesitates, and then starts coming back again. Then it is speeding up all the way until it gets to the equinox, at which time it is moving fastest. It'll displace the greatest

FIGURE 7.18. Hand-held images of the setting of the moon on the morning of the 2022 equinox, looking south from the Pilot Knob Inn.

> amount of distance along the horizon and then again, it'll be slowing down as you get to the summer solstice. So, it's this oscillation back and forth. It is a rhythm and it has been going on for a very long time and it was always very important to the ancient peoples.
>
> —Randall Carlson (2021)[41]

That appearance of the sun slowing down makes the cross quarters seem off in terms of the actual ticking off of the calendar days, but they are correct in the tracking of the sun on the landscape. The sun appears, from the perspective of earth, to slow down and speed up on the horizon. To research archeoastronomy, the general terms to know are the vernal equinox, the summer solstice, the autumnal equinox, and the winter solstice. With the cross quarters, you have the Imbolc (Imbolg), Beltane (Bealtaine), Lughnasadh (Lughnasa), and Samhain (Halloween). World-changing events often happen around these eight dates. Stones on top of the Big Pinnacle might mark the rising and setting of these symbolic times of the year or be marked on the horizon by a natural or man-made notch.

In addition, could the 18.61-year lunar declination cycle (lunar standstill or lunistice) also be revealed on the circle within a circle on the Big Pinnacle?[42] For the general public, these marked dates in the landscape are referred to in negative

41. Randall Carlson, "Kosmographia Live 003 Solstice and Precessional Motion," streamed live on December 20, 2021, YouTube video, https://www.youtube.com/watch?v=ioZ-kNov20c.
42. Prehistoric Solar Calendar, *Mega-What*, http://mega-what.com/glossary/SolarCal.html; Prehistoric Lunar Calendar, *Mega-What*, http://mega-what.com/glossary/LunarCal.html.

terms as pagan (sun worshipers, didn't have the wheel, hunter-gather, no authority figure, often matriarchal). However, they remain important farming dates and are clearly important to Masonic Freemasons as prime symbols of the year, which they build into their structures, like the great cathedrals of Europe. If you are new to gardening, almanacs feature guides and information about using the phases of the moon.

> The moon has a monthly cycle but also follows a longer pattern. Every eighteen and a half years it rises at its most northern position and then two weeks later at its most southern. This celestial event last happened in [June] 2006. It won't happen again until [April] 2025. It is called a lunar standstill.
>
> —Musician-Narrator Robbie Robertson (2018)[43]

Let's review the definition of solstice and equinox since this applies to the shadow moving east across the land near Pilot Mountain:

> Equinox descends from aequus, the Latin word for 'equal' or 'even,' and nox, the Latin word for 'night'—a fitting history for a word that describes days of the year when the daytime and nighttime are equal in length. In the northern hemisphere, the vernal equinox marks the first day of spring and occurs when the sun moves north across the equator. (Vernal comes from the Latin word ver, meaning 'spring.') The autumnal equinox marks the first day of autumn in the northern hemisphere and occurs when the sun crosses the equator going south. In contrast, a solstice is either of the two moments in the year when the sun's apparent path is farthest north or south from the equator.
>
> Solstice gets its shine from sol, the Latin word for 'sun.' The ancients added sol to -stit- (a participial stem of ister, which means 'to stand still') and came up with solstitium. Middle English speakers shortened solstitium to solstice in the 14th century.[44]

---

43. *Cities of the Sky*, episode 3, directed by Gary Glassman and Joseph Sousa, produced by Rob Tinworth, et al. (Public Broadcasting Service, 2018), 2:29, https://video.alexanderstreet.com/watch/cities-of-the-sky.
44. *Merriam-Webster.com Dictionary*, s.v. "equinox," accessed 2020, https://www.merriam-webster.com/dictionary/equinox; *Merriam-Webster.com Dictionary*, s.v. "solstice," accessed 2020, https://www.merriam-webster.com/dictionary/solstice.

VERNAL EQUINOX ON MARCH 20, 2022, REVEALS A PYRAMID SHADOW

A unique thing happens on the autumnal and vernal equinox sunset around Pilot Mountain, which adds weight to the term "natural curiosity" from the 1800s.[45] This phenomenon happens at sunrise or sunset. Unfortunately, it was missed during the daytime autumnal-equinox visit to the mountain in 2020 with the Beasley grandchildren. As the sun begins to set, the entire mountain casts a shadow, with the Big Pinnacle as a large pointer, that begins even closer to the mountain than the Pilot Mountain Visitor Center.

It is the nearly perfect east-west aspect of the Pilot Mountain that brings this out, but on-site research hasn't found a viewpoint for the sunrise shadow yet. You can do this in your own backyard (if you get full sun there) with a simple pole, or, you could build your own elaborate earthwork pyramid. If you do build a small earthwork pyramid, it should be layered with clay, gravel, and soil in a bowl shape to make it last for a few thousand years or more. This is termed "below ground assets" with all sorts of structures in the Mound Culture (e.g., ponds, walls, pyramids, and simple and complex mounds). Randall Carlson humorously advises to get a comfortable chair to watch the shadow move around the stick if you are doing it for the full year at home.[46]

This Pilot Mountain State Park pyramid shadow image was taken on the vernal equinox as the sun set on the Big Pinnacle with the camera facing east (fig. 7.19). That particular equinox took place March 20, 2022, at 11:33 a.m., which was seven hours earlier than the image taken. Given the distance from the equator, the sun was slightly off the east-west aspect of Pilot Mountain but still impressive in the line-up. On this vernal equinox, the sunrise and sunset were photographed for this book. We also observed a sun dagger on the Big Pinnacle at the moment of the equinox, so we got three vastly different points of view on that one day for the book and website. The moment of the solstice is also revealed at the same sun dagger location of the equinox, but the light appears to reach it from a small opening in the rock. While the Forsyth Astronomy Club and the UNC-CH– Planetarium director were contacted about this phenomenon, neither seemed to have an interest in archeoastronomy on Pilot Mountain. However, the Forsyth Astronomy Club does meet on Pilot Mountain periodically to view the night sky away from big city lights.

The first image taken in a series had the pyramid shadow only to the visitor center parking lot below and it was in fairly sharp detail. The Big Pinnacle shadow, as a pointer, went right over the Pilot Mountain State Park Visitor Center perfectly on the equinox. The shadow then stretched and grew with each passing moment, moving towards southeast of the Sauratown Mountains, at least five miles away and well beyond. It was as captivating as witnessing the aurora borealis or a full eclipse anywhere on the planet. We shared a magical feeling of awe while witnessing the prehistoric cosmic calendar at Pilot Mountain in modern

45. Glace, Glace, and Sperry, interview.

46. *Randall Carlson Podcast*, "Kosmographia Live 003," 16:00.

times. Hikers moving swiftly around the Pilot Knob Trail stopped in their tracks to view and discuss this huge shadow. As it was discussed, the shadow continued to move away from the mountain toward the horizon.

The pyramid aspect of the shadow was also noted in May, close to the cross-quarter time of the same year (figs. 7.20 and 7.21). The second time, the entire pyramid shadow was witnessed from the start to the finish, which had dramatically changed direction because of the position of the sun. The whole process took about an hour and a half from the start to the finish. As the sun begins to set behind the Blue Ridge Mountains, it is still broad daylight, but the shadow simply fades slowly until the gigantic pyramid disappears on the horizon miles and miles away. This truly dramatic effect is like a movie transition fade and then it is simply gone. This comparatively quick fade of the pyramid shadow also allows you to get on the other side of the Big Pinnacle in time to see the actual setting of the sun if conditions are right.

Back in March of 2022, the park rangers had a very busy equinox day, which fell on a Sunday. As darkness quickly approached on the mountain, there were about half a dozen cars in the parking lot, but everyone lingered. These cars were only allowed up after the van transport system closed for the day, so there weren't many. It was a cloudless, warm evening as the sun dipped just below the horizon almost directly west. Everyone there talked like buzzing bees about the giant

FIGURE 7.19. The pyramid shadow on the 2022 vernal equinox, taken overlooking the Pilot Mountain Visitor Center on the Pilot Knob Trail. Note that the small but proportioned Big Pinnacle shadow is at the very top of the pyramid. The pyramid shadow above was in movement for ninety minutes and then slowly faded on the horizon miles away from the mountain.

FIGURES 7.20 and 7.21. A pyramid shadow (above) almost on the 2022 summer solstice. The notch in the western landscape (below), away from the mountain, was revealed during the same 2022 summer solstice just like at Serpent Mound in southern Ohio. The sun did set on that landscape notch after the pyramid shadow had faded in the east. At Serpent Mound, on the solstice, the head of the gigantic earthwork snake points directly to the notch in the landscape on the horizon. Mostly cloudy skies prevented images the day of the solstice, but some were taken within the seventy-two-hour solstice window.

FIGURE 7.22. The sun on the very cloudy morning after the autumnal equinox of 2022 came up right on the notch of the Big Pinnacle of Pilot Mountain State Park. Did this confirm the rediscovery of this possible archeoastronomy effect on Pilot Mountain? Or, is it just a natural coincidence?

pyramid shadow, which truly had a thrilling magical quality to it, as the park ranger patiently waited in a vehicle for all to leave so they could close the gate. That was groundskeeper Bert Coleman's job for decades during the private ownership by the Beasley family and even after it became Pilot Mountain State Park. Then, a wedding photographer showed us all a video on his phone from that day of doves being released at the ceremony he had just shot with Pilot Mountain in the background. All left the mountain that evening satisfied.

SEASONS ON PILOT MOUNTAINS

Sherrie Robin Lynch-Bennett in the *Yadkin Valley Living* magazine wrote,

> Each changing season brings its own characteristics to Pilot Mountain. In the spring, the mountain takes on a contrast of colors. Shades of green bring new life to the trees, as do the blooms of the dogwood and red bud trees. By Mother's Day weekend, hues of pink and white show the beauty of the blooming rhododendron and mountain laurel. In the summer, wildflowers can be found nestled along the trails and near the river. Fall comes alive with its spectacular colors like paint dropping from a paint brush onto the trees which draws park visitors from afar. The fall season

> also draws bird lovers with its annual hawk migration. Hawks migrating south are counted in the millions by the local Audubon Society. The winter months cast a sparkle to the peak when snow and ice glisten in the sunlight. Each season has its own illuminating sunset with a kaleidoscope of color that brings peace to your soul.
>
> —Sherrie Robin Lynch-Bennett (2006)[47]

FIGURE 7.23. Pilot Mountain State Park as seen from the air in an airplane after the fire of 2021.

FIGURE 7.24. Huge wall-size warehouse mural of Pilot Mountain. Courtesy of the collection of Cleve and Deborah Harris.

47. Sherrie Robin Lynch-Bennett, a.k.a. Robin L. Bennett, *Yadkin Valley Living,* July/August 2006, 30–34.

Jim McKelvey

ABOUT THE AUTHOR

ABOUT MY RESEARCH AND WRITING CAREER THAT LED TO THE WRITING OF THIS BOOK

Yes, tooting my own horn a bit here, but it is important to explain the research and writing path for this series of books on Pilot Mountain State Park. Up until now, even some of my immediate family have not heard what is shared here for the public. This section is presented to explain the foundation of professional research and writing that has been part of my entire adult life.

I was born in Michigan but moved to Virginia when I was two years old. I spent those important formative years around NASA employees' families, who worked at Langley Research Center in Hampton Roads, VA. The Mercury Seven astronauts all seemed to live nearby, and I saw some with their children at swim meets and in the grocery stores in the area on a regular basis. Loud helicopters going back and forth to Langley Air Force Base and Fort Eustis flew right over my family home daily at 5:00 p.m. Then the astronauts would mysteriously pop up on national TV recorded in Florida, which was confusing to a six-year-old.

Our home was on high ground in the midst of wetlands off the James River a dozen or so miles from Williamsburg, VA. Rattlesnakes, water moccasin snakes in the trees, and huge sea turtles are vivid memories of childhood. The eye of Hurricane Donna went over the house and caused significant destruction. In my youth, I used to swim across from Edgar Cayce's home in Virginia Beach, VA, and the Cayce archives would become important as my first research project in college my freshman year. The Edgar Cayce "past-life readings" have taken on new meaning during each phase of my own life. For a period of time, I corresponded with Gladys Turner, who was Edgar Cayce's secretary for the majority of his fourteen thousand readings, about her Tuesday-morning remembrances of working with the gifted clairvoyant.

After returning to Michigan with the family, my first job for pay was as a paperboy for the *Detroit Free Press* with 90 daily papers and 140 Sunday papers. In high school, I was an apprentice and mentored to run a Baskin-Robbins ice cream store for two years, but then I opted for college.

With an early start to college in the summer after graduating high school, I became the manager of publicity for the Major Events committee at Eastern Michigan University in Ypsilanti, Michigan, as a volunteer. Two weeks after graduating high school, I was helping two other volunteers to book musical acts for the fall and winter season at the university. This became a daily activity along with going to classes during the summer session. They negotiated and secured contracts for concert appearances by the Peter Gabriel version of Genesis; blues legends like Muddy Waters, Johnny Winter, and James Cotton; The Temptations and the Four Tops; David Crosby and Graham Nash with the Section featuring David Lindley; Santana; Dave Mason; Lynyrd Skynyrd; Peter Frampton; America; Chicago; Seals and Crofts; Bob Seger; and Arlo Guthrie, among others. All summer long, the committee would get as many as half a dozen offers of musical acts each week. The promoter who books the act, not the artist, decides on an agreed-upon local venue. Then, and only then, legal documents are exchanged and signed. Genesis was on a large arena tour, but our team of volunteers was able to book artists into an acoustically perfect auditorium with just 1,500 seats. It was an immediate sell out. Their light rig went over the first ten rows of the small facility for that show. Handling the outrageous contract riders from the bands was amusing that summer.

At eighteen years old, I witnessed unidentified flying objects (UFO) expert, J. Allen Hynek, apologize publicly to the Washtenaw County Sherriff's Department about the "swamp gas" excuse he was forced to make by the United States Air Force. The sheriffs attended the event in the front rows of the 1,500-seat Pease Auditorium on the college campus. Hynek went on to give a presentation using slides from his archives on UFOs for three hours. The talk was based on his decades of Project Blue Book work for the United States Air Force on the subject of UFOs. It was Hynek's firm opinion given during his talk that UFOs came from another dimension rather than long space travel. That is why they are seen to seemingly blink in and then blink out of sight so often. Hynek would years later become the technical advisor of Steven Spielberg's *Close Encounters of the Third Kind*, and more recently, a TV series based on his Project Blue Book work was made with twenty episodes aired over the course of two seasons. As you read the last chapter of the next book of this series, keep Hynek's dimensional theories in the back of your mind, or be open to the suggestion. Cutting-edge physics is certainly catching up with Hynek's theory.

Years later, I would write about music because of my local promoting background at the university for the Ypsilanti Press starting in 1978. As a cub reporter, I wrote my initial concert reviews that were exactly 250 words long (one page, double spaced) about the Neil Young and Crazy Horse "Rust Never Sleeps" tour, Arlo Guthrie and Pete Seeger, Bonnie Raitt, Dave Mason, and Bob Welch. Those clips, known as tear sheets, were sent to *Variety*, an internationally known entertainment magazine in New York City that began publication in 1905, and this prompted a job offer with little effort. Complimentary tickets in the first ten rows, any expenses including mileage to the show and payment per column inch were all part of that deal with *Variety* for a monthly paycheck. A notarized press badge from *Variety* turned out to be a big deal at venues in Michigan. When you read how much a movie made anywhere, those figures come directly from *Variety* reports. Under the byline, "kelv," was published approximately two hundred times, and most of those are online now as part of the digital archives of the weekly newspaper. The one rejection of something written on assignment for *Variety* happened when Emmylou Harris was misspelled "Emmy Lou Harris." Since many tours started in Michigan, I was assigned some plumb assignments from *Variety*.

B. B. King sent a telegram invitation via Western Union to me to attend a prison concert in Jackson, Michigan, on February 20, 1981. I was so surprised by the telegram that I forgot to tip the man who drove up with it. Immediately, I accepted the invitation on the phone and began to do research on King and his prison reform opinions. This turned out to be King's thirtieth prison show, and each time he invited the press to the event. The day of the Jackson State Prison show, the reporters went deeply into the prison right along with the King band and entourage that included lawyer F. Lee Bailey and Congressman John Conyers. Before each door opened, everyone was searched, but upon going through, there would be another door. This process went on for seven or eight doors with a search done at each. Each time the door behind us would slam shut loudly and only then would the next one open.

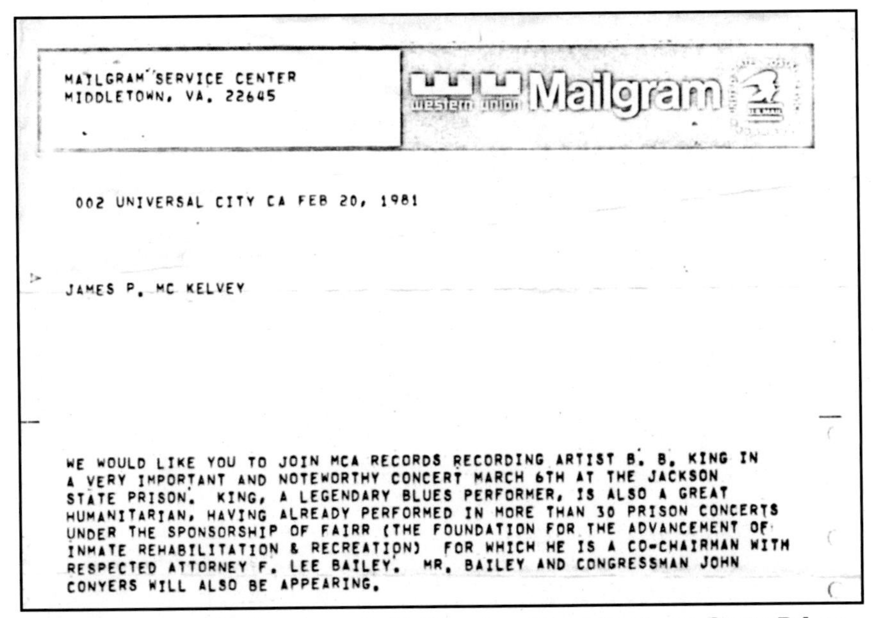
MAILGRAM SERVICE CENTER
MIDDLETOWN, VA. 22645

Western Union Mailgram

002 UNIVERSAL CITY CA FEB 20, 1981

JAMES P. MC KELVEY

WE WOULD LIKE YOU TO JOIN MCA RECORDS RECORDING ARTIST B. B. KING IN A VERY IMPORTANT AND NOTEWORTHY CONCERT MARCH 6TH AT THE JACKSON STATE PRISON. KING, A LEGENDARY BLUES PERFORMER, IS ALSO A GREAT HUMANITARIAN, HAVING ALREADY PERFORMED IN MORE THAN 30 PRISON CONCERTS UNDER THE SPONSORSHIP OF FAIRR (THE FOUNDATION FOR THE ADVANCEMENT OF INMATE REHABILITATION & RECREATION) FOR WHICH HE IS A CO-CHAIRMAN WITH RESPECTED ATTORNEY F. LEE BAILEY. MR. BAILEY AND CONGRESSMAN JOHN CONYERS WILL ALSO BE APPEARING.

FIGURES 8.1. Invitation from B. B. King to attend Jackson State Prison concert.

FIGURES 8.2. Backstage with B. B. King at Jackson State Prison. Courtesy of the BB King Music Company LLC.

King and his amazing blues band played two shows, and during the first one, I sat in the audience. Bailey took me aside to say that it was very brave to sit in the audience during the first show with only a few other reporters. All of the prisoners at the show were in prison for life with no opportunity for parole. The second show, I watched from the wings after doing a half-hour interview with King right there in the prison stage between shows.

The only front-page newspaper byline I wrote was for the *Detroit Free Press*, thanks to Neil Young. I met Neil Young's dad, Scott Young, sitting at a few shows over the years in the complimentary seats and we had become friends as fellow writers. Later we became pen pals. While talking about his dad's book that I was mentioned in, Scott Young's *Neil & Me*, Neil Young brashly told me, "I'm going to arrange an interview with Willie Nelson about Farm Aid and if you don't get it published, I'll never talk to you again." Next morning it was all arranged, and another research project was started on the formation of Farm Aid. The interview took place just a few weeks before the very first festival that raised millions for the family farm issue. That meeting turned out to be a three-hour interview on Willie Nelson's bus, sitting at the table behind the driver. It became the basis of three features in the *Detroit Free Press* the weekend of the first Farm Aid in 1985. One was on the front page of the paper below the fold, another was on the front page of the entertainment section above the fold, and third was a Q&A style in the middle of the entertainment section on the same day.

The biggest writing project before this book was when I turned my full attention to researching and writing on the life of Aretha Franklin. After speaking with all of her living immediate family members and dozens of her influential musical friends at length, I was granted an interview with Aretha Franklin for fifteen minutes. While literally walking out the door of my home for the interview, the phone rang to inform me the location had been changed from her home

FIGURE 8.4. Neil Young recognizing me at the Farm Aid press conference with a nod in my direction in Raleigh in 2014.

FIGURE 8.5. Neil Young's manager, the late Elliott Roberts, made arrangements for me to attend Farm Aid in 2014. We had a working relationship since first meeting each other at the soundboard before the show started during the "Rust Never Sleeps" tour in 1978. One of the funniest managers in the entertainment industry.

FIGURE 8.6. This image was taken the week that Aretha Franklin died. It was taken at her father's New Bethel Baptist Church in Detroit when the author paid his respects before the funeral.

in Bloomfield Hills, Michigan, to her secret apartment in downtown Detroit, Michigan. Had the phone call happened a moment later, the interview may not have happened. This was a time way before cell phones, fax machines, and email.

On the way to the interview with Aretha Franklin, the idea for an icebreaker gift surfaced since she hated talking to the press. It had been almost twenty years since she had done an interview of any substance after a rather hurtful *Time Magazine* piece on her in 1967. Just after pulling off the freeway thinking about a gift for her, a wine shop was spotted at the end of the exit across the street. The store was empty, but the manager was there, and I told him about the interview in a half hour with Aretha Franklin, who also lived in the area. The manager put up his arms and said, "I've got chill bumps, look at the hair on my arms standing up. Aretha was just here last week standing where you are standing." He immediately pointed to an image on the wall of Aretha from the week before. He said, "I waited on her and I've got just the thing for you over at the Godiva chocolate stand there behind you. I'll put that particular Godiva chocolate bar she likes in the same wrapping I did for her, and it'll only cost you $15." In less than ten minutes, I was back on the road to the interview but also wondering what the heck had just happened.

After knocking on her door, Aretha Franklin herself greeted me at the door barefoot in orange leather pants and a black t-shirt. Right at the threshold of her secret apartment in downtown Detroit, the story about the hour-long drive there and what had randomly happened at the wine shop had her full attention. She also raised up her arm and said she had goosebumps, showing the hair on her arm standing up.

We got comfortable on two plush sofas across from each other, and the interview began in her large living room using a high-end stereo microphone set-up. The Detroit River flowed swiftly right below as we spoke, since her living room was like an enclosed balcony. Every fifteen minutes from across the room, her record company rep, Traci Jordan, would say, "Times up, lets wrap up the interview!" Aretha waved her off each time, and after three hours the second ninety-minute tape ran out for the tape recorder. Then Aretha started to interview me, asking if she should do a duet with Dolly Parton and so many other topics with no tape left to record our conversation. It was suggested she buy a venue in Detroit since she had a fear of flying after a terrible incident in the air on a plane while touring, the idea being she could have people come to her rather than touring. She had even taken a fear-of-flying course, but it didn't help her. In the last months of her life, she started to explore buying a club like Buddy Guy and B. B. King had done.

That taped interview that I transcribed myself with Aretha Franklin and a review of the gospel recording session she did at her father's church in Detroit over the course of three nights with Mavis Staples ended up in *The Face* magazine in London, England. It was published on every continent in the world except Antarctica (which I am also fascinated with).

In addition to writing as a hobby, I have worked for nonprofit health institutions. I learned early in life that helping others gave me the greatest joy. At St. Joseph Mercy Hospital in Ann Arbor, Michigan, I initiated a non-smoking campaign that was successful, based on secondhand smoke research done in 1985 at the hospital medical library.

While in the healthcare field, I also managed a low-income sliding fee scale clinic in Ann Arbor and was able to write and obtain three Community Development Block Grants. Around that same time working at the low-income clinic, a single handwritten letter to the CEO of the local hospital started the "Disadvantaged Fund" for people to have medical coverage who didn't have insurance or qualify for Medicaid. Forty-five years later, that same fund is still helping people based in Ann Arbor, Michigan.

After a move south to North Carolina on the fourth day working full time at Duke University Medical Center, I raised my hand at a meeting with the new CEO asking about confidentiality there. On the spot, I was made the chairman of the Duke Confidentiality Committee and was responsible for the research to be provided to every department director before each meeting. The book, *Robert's Rules for Order*, about running a meeting became very important. The whole fifty-dollar budget for research issued that very day was used at the Duke Medical Library at five cents a copy. Thirty-five items were presented to the administration, and twenty-three were accepted by them. The phrase "Mum's the word" is still used today at the institution from that committee in 1992.

I made a move in 1992 from writing to concert photography as a hobby. The photography took place in living rooms, at house concerts, small theaters, a few arenas, and cruise ships. Many musicians liked the photographic work, so some of my freelance images done at shows ended up on official CDs by Darrell Scott, verify voter registrations or signatures on petitions.

Where does that put the Grand Old Party? Between a rock and a hard spot?

If I were a Republican candidate, I'd be calling for Mel Larsen's resignation. Unless, of course, they're scared to death of tax crusader Robert Tisch and his party.

LUCILLE DIEHL
Allen Park

IN YOUR ARTICLE, "Pierce doctors faithless voters" (Free Press, July 24), about Dr. Ed Pierce, you inaccurately implied that the Summit Medical Center in Ann Arbor is no longer open.

In 1978, when Dr. Pierce won the state Senate seat and left Summit Medical Center, he was replaced by Marlis Pacifico, MD, who remains at the center today.

The Summit Medical Center is still guided by a board of directors and fees are based on a sliding scale according to both income and the number of people supported by that income. We continue to receive private donations from the community, which are used for capital expenditures and to help offset operational deficits due to the spiraling costs of medical supplies and equipment.

JAMES P. McKELVEY
Manager
Summit Medical Center
Ann Arbor

IN THE JULY 30 editorial of best choices for state representative, the Free Press included the name of 31st District Rep. Lucille McCollough. She was given credit for her longevity, but was said to be lacking in accomplishments while in office.

A perfect attendance and a voting record has to be earned. She is the only legislator in the nation to have accomplished this.

Her record as a legislator is impressive and appreciated by the majority of voters.

LOIS ESSELTINE
Dearborn

FIGURE 8.7. Letter to the Editor of the Detroit Free Press to correct a feature article about Summit Medical Center still being open for business.

John Cowan, and Alice Gerrard, alongside folk stalwarts Jim Watson, Cliff Hale, and Chris Brashear as the Piedmont Melody Makers. Emmylou Harris linked from her homepage a series of images of her Sweet Harmony Traveling Review tour that had over one million hits online. Over 4,500 images were donated to the Southern Folklife Collection at the Louis Round Wilson Library at UNC-CH. They included 160 concerts and annual multiday festivals like Cayamo (2008–2020), Merlefest (1993–2015), and the International Bluegrass Association (2013–2020) festivals in Raleigh.

Early in 2013, I retired on a small pension from Duke University Medical Center after twenty-one years of service with the organization. That is when research began in earnest about Pilot Mountain, but it would be another seven years before the first draft for this book was started. That first draft of the book was rejected by MacFarland Publishing in Jefferson, North Carolina, but they saw something in it that made them want me to write the one you see here with footnotes that reference things that happened centuries ago. Their response was both a rejection letter and assignment letter, which made all the difference for this book today. Unfortunately, because of a long lead time to publication of at least eighteen to twenty-four months, a decision was made to start a publishing company for this series of books.

FIGURE 8.8. Wilson Library on the campus of UNC-CH, where the "James McKelvey Collection" of photography is housed.

FIGURE 8.9. The three CD covers done by the author: Darrell Scott "Live in NC"; John Cowan "8,745 ft."; and The Piedmont Melody Makers "Wonderful World Outside."

FIGURE 8.13. At times driving by it in your car, it almost looks like there must be a tunnel ahead to get past Pilot Mountain State Park as it looms so large in front of your eyes.

EPILOGUE: RESEARCH CATCH-22

This epilogue is a cautionary tale for any writer who is using North Carolina as a subject. In a curious twist of fate, just as the TIPS Technical Publishing team in Carrboro, North Carolina, was completing the final layout and index of this very book, an innocent gesture by the author almost skuttled it all. Permission to publish half the images in the book and the unique cover were almost lost.

I encountered a series of unforeseen technical legal hurdles governing park lands and property, but it didn't change my appreciation for this mysterious mountain. It has been a joy putting this research together in a unique manner that I hope honors our ancestors with their highlighted, unfiltered quotes. This information directly from our ancestors informs the public of unique aspects of this national natural landmark all but forgotten in modern times. I would encourage people to be careful on the mountain because it remains very undeveloped on purpose. Take photographs but leave only footprints on the trails and remember it is vital to not use paints, inks, or chemicals that pollute the mineral springs that flow by so many communities on the way to the ocean. Please also show respect and honor in all that you do on the mountain, which is an active shared sacred landscape to the Indigenous tribes of North Carolina.

The revelation of this "research catch-22" started innocently enough on January 1, 2023, during shooting for the cover of the book, but the issue had literally been going on for years. On New Year's Eve, the weather report showed dense fog throughout the following morning, so I got up a 5:30 a.m. and drove to the mountain. The hope was to get above the fog at Pilot Mountain State Park and take a portrait-oriented, unique but familiar image for the cover. Within minutes of arriving at the Little Pinnacle various versions of the cover shot used for this book were completed (with and without clouds). Had I known about the permitting process, the cover would not exist since it was a holiday with no time to get it approved, signed by all parties, and paid for.

At a particular spot on the Jasmine Trail, looking east you get a unique view of the Sauratown Mountains, the Big Pinnacle, and the pyramid shape of the mountain. Because of the heavy fog with quarter-of-a-mile visibility throughout the Piedmont, very few people were in the park at that hour. This spot on the Jasmine Trail is about thirty feet directly below the spot that the former owner of the mountain, J. W. Beasley, enjoyed, which earlier in the book was described as the "Beasley Overlook."

My intent was to get the image of the fog and that view J. W. Beasley liked for his grandchildren to ring in the new year. I walked the entire length of the Jasmine Trail, which is rather steep, and saw a man standing where I wanted to take the image. As I got closer and closer down the steep path, I could see the man had a park ranger uniform on. It turns out the ranger was the new Superintendent of Pilot Mountain State Park, Jason Anthony. As we introduced ourselves, I told him, "I can't think of a better spot on the mountain to meet even if we had arranged this in advance." There was a sign behind us stating, "End of trail."

We had a great conversation there for half an hour, and I think the Pilot Mountain State Park is in good hands with him. During the relaxed conversation, I mentioned I was doing a book and that his approval was needed for it to be sold in the park. He agreed to accept a copy and review it but said the bookstore staff would also have to decide to carry it since they are a private concessionaire that also owns all the inventory. Right then, we observed a fire truck quietly above us on that last hairpin turn before the parking lot. Superintendent Anthony had his radio on him but did not know why they were on the summit, so he made a pleasant but quick exit to find out what was going on.

Months later, on the morning of the vernal equinox, March 20, 2023, I left home long before sunrise to film the sun rising on the notch in the Big Pinnacle on a clear cloudless morning. It had been very windy, and the temperature was in the low twenties during the filming on the Little Pinnacle facing east. After that I went down to the Pilot Mountain Visitor Center parking lot, but it was still before the building opened.

Ranger Williams was getting out of a white truck, so I stopped her to ask if she could get a copy to Superintendent Anthony, since I had recently received them and had not yet given one to him. I told her to feel free to share the cover with the other park rangers as I was rather happy with it. Ranger Williams said she was new and would do that for me. I also gave her one of my calling cards to give to Superintendent Anthony, since I had just had them made up the month before and hadn't had one when I met him on New Year's Day.

The calling card brought unexpected legal hurdles into focus. It turns out that giving a calling card as a writer who is going to be publishing a book is a violation of state law on state park property. Giving a calling card can be considered both solicitation for a business and also advertising. The following day around supper time, I received a call from Ranger Williams, who explained in detail over the course of ten minutes what I had been doing wrong for years and not just on the equinox when we met.

The calling card turned out to be one of many unforeseen technical legal hurdles for me. One must remember that the park rangers are only enforcing state law on the property. They did not make the rules but must enforce them. I call my difficulties a strange twist of fate because both the North Carolina State Parks and the North Carolina Office of State Archeology are also restrained. While they enforce different sets of federal, state, and general statutes, their particular technical legal hurdles impact the sharing of archeological information with the public. To be clear, nearly all the relevant published peer-reviewed archeology reports, artifacts, relics, and features on Pilot Mountain State Park plus all of Surry County cannot be shared with either the public or the area museums.

One archeology report was written in 1983 all about Pilot Mountain State Park. That report is two hundred pages long and, to this day, forty years later, has restricted access due to these governing laws that the North Carolina Office of State Archeology must follow. Another example is a showcase at the Pilot Mountain Visitor Center sponsored by Wake Forest University with artifacts from the surrounding counties. A careful analysis will reveal none of the artifacts in that showcase are from Surry County or Pilot Mountain State Park.

In my case for this book, the following laws apply, but we will delve deeply into the laws for the North Carolina Office of Archeology in the next book that is nearly complete:

> Section .1100 - Commercial Enterprises: Advertising: Meetings: Exhibitions: ETC.
>
> 07 NCAFC 13 B .11.1 Commercial Enterprises
>
> Only Park employees, contractors, or their agents may engage in business or conduct commercial activity in a park unless authorized by a permit as provided by Rule .0104 of this Subchapter.

> 07 NCAC 13B .1102 Commercial Photography, Filming and Recording
>
> A person shall not photograph, film, or make other recordings within any park for commercial purposes unless the person has a Special Activity Permit. Applications for permits may be made as provided by Rule .0104 of this Subchapter.
>
> 07 NCAC 13B .1104 Advertising
>
> No one except authorized park employees, their agents, or contractors may erect or post within any park a notice or advertisement without first obtaining a permit. Applications for permits may be made as provided by Rule .0104 of this Subchapter.

I have been advised that because this has gone from being a hobby and has evolved into a business, I must observe the laws that the rangers must enforce that would apply to any business in terms of my conduct at a state park. This means I can no longer talk to the rangers or public about the book while on the property unless I have an approved Special Activity Permit signed by all parties in advance and have paid a fee.

Unless I have a permit, I can't acknowledge the book on the property. Since clearly this could cause some very awkward conversations on the mountain between myself and the public, I apologize in advance. I have received a verbal warning about my conduct researching the book at Pilot Mountain State Park.

Each state park enforces a permit process, which includes fees. The three categories of permits are Scouting, DPR Special Activity Permit Application, and DNCR Film Permit Application. Permits must be signed by all parties before they can be approved and fees paid. One wonders how many permits were given out when the fire on the entire mountain became national news and the national, state, regional, and local media covered it daily for over a week.

If you are reading this book now then some sort of compromise was met between the author and Pilot Mountain State Park for use of the filming and photography.

> Most of the requests we get for photography and filming are for commercial photo shoots (engagement pictures, graduations, etc.) and the occasional indie film. We have never had an author request to photograph in the park for a book. A few works of fiction have been written about Pilot Mountain, but it is used as a setting only which uses just a generic image of the mountain that the author could have gotten anywhere."
>
> —Superintendent Jason Anthony (2023)[1]

1. Superintendent Jason Anthony, official email, March 27, 2023, 5:10 p.m.

The following quote by the Dogon tribe in Mali seems so appropriate here for this epilogue. It is also the way Canadian author Shannon Dorey ended her book *The Nummo: The Truth About Human Origins*.[2] If you like the writing of Joseph Campbell, you would really appreciate Shannon Dorey's series of three books on the Dogon tribe.

As one aspect of the next book on Pilot Mountain State Park, we will explore the many similarities of the Dogon (Nummo) and the Cherokee (Nunne'hi) immortals in their oral tradition stories. Water and fire play a vital role in the oral tradition of the immortals of both the Dogon and Cherokee. I am now fascinated by how Indigenous oral history stories across the entire planet hang together so beautifully but have been suppressed so severely by the political elite for nearly all of human history.

> What is left—which is unknown—will be known later to man [humans] and will change the world. It is said that this revelation will come slowly at first like a fog, then swiftly like the rain and wind.[3]

FIGURE 9.2. Image taken the clear day before the autominal equinox of 2022.

2. Shannon Dorey, *The Nummo: The Truth about Human Origins* (Elemental Expressions Ltd, 2004, with revisions 2006, 2008, 2013, and 2019), 340.
3. Marcel Griaule and Germaine Dieterlen, *The Pale Fox*, trans. Stephen C. Infantino (Chino Valley, Arizona: Continuum Foundation, 1986), 449.

FIGURE 9.3. Sunrise on the Big Pinnacle, looking due east from the Little Pinnacle, on March 20, 2023

INDEX

C

H

I

J

K

L

M

N

O

P

Q

R

S

T

U

V

W

Y

Sunrise May 1, 2023, on the Beltane cross-quarter right between the equinox and the solstice revealed the sun over the Sauratown Mountains from the viewing area of the Little Pinnacle. Often considered the most important day of the year by ancient peoples going back millennia, this is a time-honored day for sprouting of plants in the field and putting flowers on your door (female symbol) plus a time of maypoles (male symbol) and the crowning of the May Queen. In ancient times, courtship often started on May 1, and then six weeks later June weddings were traditional. It should be noted this May Day holiday celebrated in over sixty-six countries has nothing to do with the distress call, mayday, mayday, mayday.